ASE Test Preparation Series

Automobile Test

Exhaust Systems (Test X1)

4th Edition

Australia • Brazil • Japan • Korea • Mexico • Singapore • Spain • United Kingdom • United States

ASE Test Preparation Series: Automobile Test: Exhaust Systems (Test X1), Fourth Edition

Vice President,Technology Professional Business Unit: Gregory L. Clayton

Product Development Manager: Kristen Davis

Product Manager: Kim Blakey

Editorial Assistant: Vanessa Carlson

Director of Marketing: Beth A. Lutz

Marketing Specialist: Brian McGrath

Marketing Coordinator: Marissa Maiella

Production Manager: Andrew Crouth

Production Editor: Kara A. DiCaterino

Senior Project Editor: Christopher Chien

XML Architect: Jean Kaplansky

Cover Design: Michael Egan

Cover Images: Portion courtesy of DaimlerChrysler Corporation

ISBN-13: 978-1-4180-3886-1

ISBN-10: 1-4180-3886-5

Delmar
Executive Woods
5 Maxwell Drive
Clifton Park, NY 12065
USA

Cengage Learning is a leading provider of customized learning solutions with office locations around the globe, including Singapore, the United Kingdom, Australia, Mexico, Brazil, and Japan. Locate your local office at: **international.cengage.com/region**

Cengage Learning products are represented in Canada by Nelson Education, Ltd.

For your lifelong learning solutions, visit **delmar.cengage.com**

Visit our corporate website at **www.cengage.com**

Printed in the United States of America
2 3 4 5 6 13 12 11 10 09

ED301

Contents

Preface v

Section 1 The History and Purpose of ASE

Section 2 Take and Pass Every ASE Test

How are the tests administered? 3
Who Writes the Questions? 3
Objective Tests 4
Preparing for the Exam 5
During the Test 5
Testing Time Length 6
Your Test Results! 6

Section 3 Types of Questions on an ASE Exam

Multiple-Choice Questions. 9
EXCEPT Questions 10
Technician A, Technician B Questions 10
Most-Likely Questions. 11
LEAST-Likely Questions. 12
Summary. 12

Section 4 Overview of the Task List

Exhaust Systems (Test X1) 13
Task List and Overview 14
A. Exhaust System Inspection and Repair (11 Questions). 14
B. Emission Systems Diagnosis (8 Questions) 26
C. Exhaust System Fabrication (6 Questions). 34
D. Exhaust System Installation (8 Questions). 40
E. Exhaust System Repair Regulations (7 Questions) 46

Section 5 Sample Test for Practice

Sample Test 53

Section 6 Additional Test Questions for Practice

Additional Test Questions 65

Section 7 Appendices

Answers to the Test Questions for the Sample Test Section 5 77
Explanations to the Answers for the Sample Test Section 5 78
Answers to the Test Questions for the Additional Test Questions Section 6 91
Explanations to the Answers for the Additional Test Questions Section 6 92
Glossary 107

Preface

Delmar Learning is very pleased that you have chosen our ASE Test Preparation Series to prepare yourself for the automotive ASE Examination. These guides are available for all of the automotive areas including A1–A8, the L1 Advanced Diagnostic Certification, the P2 Parts Specialist, the C1 Service Consultant and the X1 Undercar Specialist. These guides are designed to introduce you to the Task List for the test you are preparing to take, give you an understanding of what you are expected to be able to do in each task, and take you through sample test questions formatted in the same way the ASE tests are structured.

If you have a basic working knowledge of the discipline you are testing for, you will find Delmar Learning's ASE Test Preparation Series to be an excellent way to understand the "must know" items to pass the test. These books are not textbooks. Their objective is to prepare the technician who has the requisite experience and schooling to challenge ASE testing. It cannot replace the hands-on experience or the theoretical knowledge required by ASE to master vehicle repair technology. If you are unable to understand more than a few of the questions and their explanations in this book, it could be that you require either more shop-floor experience or further study. Some resources that can assist you with further study are listed on the rear cover of this book.

Each book begins with an item-by-item overview of the ASE Task List with explanations of the minimum knowledge you must possess to answer questions related to the task. Following that there are 2 sets of sample questions followed by an answer key to each test and an explanation of the answers to each question. A few of the questions are not strictly ASE format but were included because they help teach a critical concept that will appear on the test. We suggest that you read the complete Task List Overview before taking the first sample test. After taking the first test, score yourself and read the explanation to any questions that you were not sure about, including the questions you answered correctly. Each test question has a reference back to the related task or tasks that it covers. This will help you to go back and read over any area of the task list that you are having trouble with. Once you are satisfied that you have all of your questions answered from the first sample test, take the additional tests and check them. If you pass these tests, you will be prepared to do well on the ASE test.

Our Commitment to Excellence

The 4th edition of Delmar Learning's ASE Test Preparation Series has been through a major revision with extensive updates to the ASE's task lists, test questions, and answers and explanations. Delmar Learning has sought out the best technicians in the country to help with the updating and revision of each of the books in the series.

About the Series Advisor

To promote consistency throughout the series, a series advisor took on the task of reading, editing, and helping each of our experts give each book the highest level of accuracy possible. Dan Perrin has served in the role of Series Advisor for the 4th edition of the ASE Test Preparation Series. Dan began ASE testing with the first series of tests in 1972 and has been continually certified ever since. He holds ASE master status in automotive, truck, collision, and machinist. He is also L1, L2, and alternated fuels certified, along with some others that have expired. He has been an automotive educator since 1979, having taught at the secondary, post-secondary, and industry levels. His service includes participation on boards that include the North American Council of Automotive Teachers (NACAT), the Automotive Industry Planning Council (AIPC), and the National Automotive Technicians Education Foundation (NATEF). Dan currently serves as the Executive Manager of NACAT and Director of the NACAT Education Foundation.

Thanks for choosing Delmar Learning's ASE Test Preparation Series. All of the writers, editors, Delmar Staff, and myself have worked very hard to make this series second to none. I know you are going to find this ook accurate and easy to work with. It is our o ective to constantly improve our product at Delmar y responding to feed ack.

If you have any uestions concerning the ooks in this series, you can email me at autoexpert@trainingbay.com.

Dan Perrin
Series Advisor

1 The History and Purpose of ASE

ASE began as the National Institute for Automotive Service Excellence (NIASE). It was founded as a non-profit independent entity in 1972 by a group of industry leaders with the single goal of providing a means for consumers to distinguish between incompetent and competent technicians. It accomplishes this goal by testing and certification of repair and service professionals. From this beginning it has evolved to be known simply as ASE (Automotive Service Excellence) and today offers more than 40 certification exams in automotive, medium/heavy duty truck, collision, engine machinist, school bus, parts specialist, automobile service consultant, and other industry-related areas. At this time there are more than 400,000 professionals with current ASE certifications. These professionals are employed by new car and truck dealerships, independent garages, fleets, service stations, franchised service facilities, and more. ASE continues its mission by also providing information that helps consumers identify repair facilities that employ certified professionals through its Blue Seal of Excellence Recognition Program. Shops that have a minimum of 75% of their repair technicians ASE certified and meet other criteria can apply for and receive the Blue Seal of Excellence Recognition from ASE.

ASE recognized that educational programs serving the service and repair industry also needed a way to be recognized as having the faculty, facilities, and equipment to provide a quality education to students wanting to become service professionals. Through the combined efforts of ASE, industry, and education leaders, the non-profit National Automotive Technicians Education Foundation (NATEF) was created to evaluate and recognize training programs. Today more than 2000 programs are ASE certified under the standards set by the service industry. ASE/NATEF also has a certification of industry (factory) training program known as CASE. CASE stands for Continuing Automotive Service Education and recognizes training provided by replacement parts manufacturers as well as vehicle manufacturers.

ASE certification testing is administered by the American College Testing (ACT). Strict standards of security and supervision at the test centers insure that the technician who holds the certification earned it. Additionally ASE certification also requires that the person passing the test to be able to demonstrate that they have two years of work experience in the field before they can be certified. Test questions are developed by industry experts that are actually working in the field being tested. There is more detail on how the test is developed and administered in the next section. Paper and pencil tests are administered twice a year at over seven hundred locations in the United States. Computer based testing is now also available with the benefit of instant test results at certain established test centers. The certification is valid for five years and can be recertified by retesting. So that consumers can recognize certified technicians, ASE issues a jacket patch, certificate, and wallet card to certified technicians and makes signs available to facilities that employ ASE certified technicians.

You can contact ASE at any of the following:

National Institute for Automotive Service Excellence
101 Blue Seal Drive S.E.
Suite 101
Leesburg, VA 20175
Telephone 703-669-6600
FAX 703-669-6123
www.ase.com

2 Take and Pass Every ASE Test

Participating in an Automotive Service Excellence (ASE) voluntary certification program gives you a chance to show your customers that you have the "know-how" needed to work on today's modern vehicles. The ASE certification tests allow you to compare your skills and knowledge to the automotive service industry's standards for each specialty area.

If you are the "average" automotive technician taking this test, you are in your mid-thirties and have not attended school for about fifteen years. That means you probably have not taken a test in many years. Some of you, on the other hand, have attended college or taken postsecondary education courses and may be more familiar with taking tests and with test-taking strategies. There is, however, a difference in the ASE test you are preparing to take and the educational tests you may be accustomed to.

How are the tests administered?

ASE test are administered at over 750 test sites in local communities. Paper and pencil tests are the type most widely available to technicians. Each tester is given a booklet containing questions with charts and diagrams where required. You can mark in this test booklet but no information entered in the booklet is scored. Answers are recorded on a separate answer sheet. You will enter your answers, using a number 2 pencil only. ASE recommends you bring four sharpened number 2 pencils that have erasers. Answer choices are recorded by coloring in the blocks on the answer sheet. The answer sheets are scanned electronically and the answers tabulated. For test security, test booklets include randomly generated questions. Your answer key must be matched to the proper booklet so it is important to correctly enter the booklet serial number on the answer sheet. All instructions are printed on the test materials and should be followed carefully.

ASE has introduced Computer Based Testing (CBT) at some locations. While the test content is the same for both testing methods the CBT tests have some unique requirements and advantages. It is strongly recommended that technicians considering the CBT tests go the ASE web page at www.ASE.com and review the conditions and requirements for this type of test. There is a demonstration of a CBT that allows you to experience this type of test before you register. Some technicians find this style of testing provides an advantage, while others find operating the computer a distraction. One significant benefit of CBT is the availability of instant results. You can receive your test results before you leave the test center. CBT testing also offers increased flexibility in scheduling. The cost for taking CBTs is slightly higher than paper and pencil tests and the number of testing sites is limited. The first time test taker may be more comfortable with the paper and pencil tests but technicians now have a choice.

Who Writes the Questions?

The questions are written by service industry experts in the area being tested. Each area will have its own technical experts. Questions are entirely job related. They are designed to test the skills you need to be a successful technician. Theoretical knowledge is important and necessary to answer the questions, but the ability to apply that knowledge is the basis of ASE test questions.

Each question has its roots in an ASE "item-writing" workshop where service representatives from automobile manufacturers (domestic and import), aftermarket parts and equipment manufacturers, working technicians, and vocational educators meet in a workshop setting to share ideas and translate them into test questions. Each test question written by these experts must survive review by all members of the group.

The questions are written to deal with practical application of soft skills and system knowledge experienced by technicians in their day-to-day work.

All questions are pre-tested and quality-checked on a national sample of technicians. Those questions that meet ASE standards of quality and accuracy are included in the scored sections of the tests; the "rejects" are sent back to the drawing board or discarded altogether.

Each certification test is made up of between forty and eighty multiple-choice questions.

Note: Each test could contain additional questions that are included for statistical research purposes only. Your answers to these questions will not affect your score, but since you do not know which ones they are, you should answer all questions on the test. The five-year Recertification Test will cover the same content areas as those listed above. However, the number of questions in each content area of the Recertification Test will be reduced by about one-half.

Objective Tests

A test is called an objective test if the same standards and conditions apply to everyone taking the test and there is only one correct answer to each question.

Objective tests primarily measure your ability to recall information. A well-designed objective test can also test your ability to understand, analyze, interpret, and apply your knowledge. Objective tests include true-false, multiple choice, fill in the blank, and matching questions. ASE's tests consist exclusively of four-part multiple-choice objective questions.

The following are some strategies that may be applied to your tests.

Before beginning to take an objective test, quickly look over the test to determine the number of questions, but do not try to read through all of the questions. In an ASE test, there are usually between forty and eighty questions, depending on the subject. Read through each question before marking your answer. Answer the questions in the order they appear on the test. Leave the questions blank that you are not sure of and move on to the next question. You can return to those unanswered questions after you have finished the others. They may be easier to answer at a later time after your mind has had additional time to consider them on a subconscious level. In addition, you might find information in other questions that will help you recall the answers to some of them.

Do not be obsessed by the apparent pattern of responses. For example, do not be influenced by a pattern like **D, C, B, A, D, C, B, A** on an ASE test.

There is also a lot of folk wisdom about taking objective tests. For example, there are those who would advise you to avoid response options that use certain words such as *all, none, always, never, must,* and *only,* to name a few. This, they claim, is because nothing in life is exclusive. They would advise you to choose response options that use words that allow for some exception, such as *sometimes, frequently, rarely, often, usually, seldom,* and *normally.* They would also advise you to avoid the first and last option (A and D) because test writers, they feel, are more comfortable if they put the correct answer in the middle (B and C) of the choices. Another recommendation often offered is to select the option that is either shorter or longer than the other three choices because it is more likely to be correct. Some would advise you to never change an answer since your first intuition is usually correct.

Although there may be a grain of truth in this folk wisdom, ASE test writers try to avoid them and so should you. There are just as many **A** answers as there are **B** answers, just as many **D** answers as **C** answers. As a matter of fact, ASE tries to balance the answers at about 25 percent per choice **A, B, C,** and **D.** There is no intention to use "tricky" words, such as outlined above. Put no credence in the opposing words "sometimes" and "never," for example.

Multiple-choice tests are sometimes challenging because there are often several choices that may seem possible, and it may be difficult to decide on the correct choice. The best strategy, in this case, is to first determine the correct answer before looking at the options. If you see the answer you decided on, you should still examine the options to make sure that none seem more correct than yours. If you do not know or are not sure of the answer, read each option very carefully and try to eliminate those options that you know to be wrong. That way, you can often arrive at the correct choice through a process of elimination.

If you have gone through all of the test and you still do not know the answer to some of the questions, then guess. Yes, guess. You then have at least a 25 percent chance of being correct. If you leave the question blank, you have no chance. Your score is based on the number of questions answered correctly.

Preparing for the Exam

The main reason we have included so many sample and practice questions in this guide is, simply, to help you learn what you know and what you don't know. We recommend that you work your way through each question in this book. Before doing this, carefully look through Section 3; it contains a description and explanation of the question types you'll find on an ASE exam.

Once you understand what the questions will look like, move to the sample test. Answer one of the sample questions (Section 5) then read the explanation (Section 7) to the answer for that question. If you don't feel you understand the reasoning for the correct answer, go back and read the overview (Section 4) for the task that is related to that question. If you still don't feel you have a solid understanding of the material, identify a good source of information on the topic, such as a textbook, and do some more studying.

After you have completed all of the sample test items and reviewed your answers, move to the additional questions (Section 6). This time answer the questions as if you were taking an actual test. Do not use any reference or allow any interruptions in order to get a feel for how you will do on an actual test. Once you have answered all of the questions, grade your results using the answer key in Section 7. For every question that you gave a wrong answer to, study the explanations to the answers and/or the overview of the related task areas. Try to determine the root cause for your missing the question. The easiest thing to correct is learning the correct technical content. The hardest thing to correct is behaviors that lead you to a wrong conclusion. If you knew the information but still got it wrong there is a behavior problem that will need to be corrected. An example would be reading too quickly and skipping over words that affect your reasoning. If you can identify what you did that caused you to answer the question incorrectly you can eliminate that cause and improve your score. Here are some basic guidelines to follow while preparing for the exam:

- Focus your studies on those areas you are weak in.
- Be honest with yourself while determining if you understand something.
- Study often but in short periods of time.
- Remove yourself from all distractions while studying.
- Keep in mind the goal of studying is not just to pass the exam, the real goal is to learn!
- Prepare physically by getting a good night's rest before the test and eat meals that provide energy but do not cause discomfort.
- Arrive early to the test site to avoid long waits as test candidates check in and to allow all of the time available for your tests.

During the Test

On paper and pencil tests you will be placing your answers on a sheet where you will be required to color in your answer choice. Stray marks or incomplete erasures may be picked up as an answer by the electronic reader, so be sure only your answers end up on the sheet. One of the biggest problems an adult faces in test taking, it seems, is placing the answer in the correct spot on the answer sheet. Make certain that you mark your answer for, say, question 21, in the space on the answer sheet designated for the answer for question 21. A correct response in the wrong line will probably result in two questions being marked wrong, one with two answers (which could include a correct answer but will be scored wrong) and the other with no answer. Remember, the answer sheet on the written test is machine scored and can only "read" what you have colored in.

If you finish answering all of the questions on a test and have remaining time, go back and review the answers to those questions that you were not sure of. You can often catch careless errors by using the remaining time to review your answers. Carefully check your answer sheet for blank answer blocks or missing information.

At practically every test, some technicians will invariably finish ahead of time and turn their papers in long before the final call. Some technicians may be doing recertification tests and others may be taking fewer tests than you. Do not let them distract or intimidate you.

It is not wise to use less than the total amount of time that you are allotted for a test. If there are any doubts, take the time for review. Any product can usually be made better with some additional effort. A test is no exception. It is not necessary to turn in your test paper until you are told to do so.

Testing Time Length

An ASE written test session is four hours. You may attempt from one to a maximum of four tests in one session. It is recommended, however, that no more than a total of 225 questions be attempted at any test session. This will allow for just over one minute for each question.

Visitors are not permitted at any time. If you wish to leave the test room, for any reason, you must first ask permission. If you finish your test early and wish to leave, you are permitted to do so only during specified dismissal periods.

You should monitor your progress and set an arbitrary limit to how much time you will need for each question. This should be based on the number of questions you are attempting. It is suggested that you wear a watch because some facilities may not have a clock visible to all areas of the room.

Computer-Based Tests are allotted a testing time according to the number of questions ranging from one half hour to one and one half hours. Advanced level tests are allowed two hours. This time is by appointment and you should be sure to be on time to insure that you have all of the time allocated. If you arrive late for a CBT test appointment you will only have the amount of time remaining on your appointment.

Your Test Results!

You can gain a better perspective about tests if you know and understand how they are scored. ASE's tests are scored by American College Testing (ACT), a nonpartial, unbiased organization having no vested interest in ASE or in the automotive industry.

Each question carries the same weight as any other question. For example, if there are fifty questions, each is worth 2 percent of the total score. The passing grade is 70 percent. That means you must correctly answer thirty-five of the fifty questions to pass the test.

The test results can tell you:

- where your knowledge equals or exceeds that needed for competent performance, or
- where you might need more preparation.

Your ASE test score report is divided into content areas and will show the number of questions in each content area and how many of your answers were correct. These numbers provide information about your performance in each area of the test. However, because there may be a different number of questions in each content area of the test, a high percentage of correct answers in an area with few questions may not offset a low percentage in an area with many questions.

It should be noted that one does not "fail" an ASE test. The technician who does not pass is simply told "More Preparation Needed." Though large differences in percentages may indicate problem areas, it is important to consider how many questions were asked in each area. Since each test evaluates all phases of the work involved in a service specialty, you should be prepared in each area. A low score in one area could keep you from passing an entire test.

There is no such thing as average. You cannot determine your overall test score by adding the percentages given for each task area and dividing by the number of areas. It doesn't work that way because there generally are not the same number of questions in each task area. A task area with twenty questions, for example, counts more toward your total score than a task area with ten questions.

Your test report should give you a good picture of your results and a better understanding of your strengths and weaknesses for each task area.

If you fail to pass the test, you may take it again at any time it is scheduled to be administered. You are the only one who will receive your test score. Test scores will not be given over the telephone by ASE nor will they be released to anyone without your written permission.

3 Types of Questions on an ASE Exam

ASE certification tests are often thought of as being tricky. They may seem to be tricky if you do not completely understand what is being asked. The following examples will help you recognize certain types of ASE questions and avoid common errors.

Paper-and-pencil tests and computer-based test questions are identical in content and difficulty. Most initial certification tests are made up of forty to eighty multiple-choice questions. Multiple-choice questions are an efficient way to test knowledge. To answer them correctly, you must think about each choice as a possibility, and then choose the one that best answers the question. To do this, read each word of the question carefully. Do not assume you know what the question is about until you have finished reading it.

About 10 percent of the questions on an actual ASE exam will use an illustration. These drawings contain the information needed to correctly answer the question. The illustration must be studied carefully before attempting to answer the question. Often, technicians look at the possible answers then try to match up the answers with the drawing. Always do the opposite; match the drawing to the answers. When the illustration is showing an electrical schematic or another system in detail, look over the system and try to figure out how the system works before you look at the question and the possible answers.

Multiple-Choice Questions

The most common type of question used on ASE Tests is the multiple-choice question. This type of question contains three "distracters" (wrong answers) and one "key" (correct answer). When the questions are written effort is made to make the distracters plausible to draw an inexperienced technician to one of them. This type of question gives a clear indication of the technician's knowledge. Using multiple criteria including cross-sections by age, race, and other background information, ASE is able to guarantee that a question does not bias for or against any particular group. A question that shows bias toward any particular group is discarded. If you encounter a question that you are unsure of, reverse engineer it by eliminating the items that it cannot be. For example:

A rocker panel is a structural member of which vehicle construction type?

A. Front-wheel drive
B. Pickup truck
C. Unibody
D. Full-frame

Analysis:

This question asks for a specific answer. By carefully reading the question, you will find that it asks for a construction type that uses the rocker panel as a structural part of the vehicle.

Answer A is wrong. Front-wheel drive is not a vehicle construction type.
Answer B is wrong. A pickup truck is not a type of vehicle construction.
Answer C is correct. Unibody design creates structural integrity by

welding parts together, such as the rocker panels, but does not require exterior cosmetic panels installed for full strength.
Therefore, the correct answer is C. If the question was read quickly and the words "construction type" were passed over, answer A may have been selected.
Answer D is wrong. Full-frame describes a body-over-frame construction type that relies on the frame assembly for structural integrity.
Therefore, the correct answer is C. If the question was read quickly and the words "construction type" were passed over, answer A may have been selected.

EXCEPT Questions

Another type of question used on ASE tests has answers that are all correct except one. The correct answer for this type of question is the answer that is wrong. The word "**EXCEPT**" will always be in capital letters. You must identify which of the choices is the wrong answer. If you read quickly through the question, you may overlook what the question is asking and answer the question with the first correct statement. This will make your answer wrong. An example of this type of question and the analysis is as follows:

All of the following are tools for the analysis of structural damage **EXCEPT:**

A. height gauge
B. tape measure.
C. dial indicator.
D. tram gauge.

Analysis:

The question really requires you to identify the tool that is not used for analyzing structural damage. All tools given in the choices are used for analyzing structural damage except one. This question presents two basic problems for the test-taker who reads through the question too quickly. It may be possible to read over the word "**EXCEPT**" in the question or not think about which type of damage analysis would use answer C. In either case, the correct answer may not be selected. To correctly answer this question, you should know what tools are used for the analysis of structural damage. If you cannot immediately recognize the incorrect tool, you should be able to identify it by analyzing the other choices.

Answer A is wrong. A height gauge may be used to analyze structural damage.
Answer B is wrong. A tape measure may be used to analyze structural damage.
Answer C is correct. A dial indicator may be used as a damage analysis tool for moving parts, such as wheels, wheel hubs, and axle shafts, but would not be used to measure structural damage.
Answer D is wrong. A tram gauge is used to measure structural damage.

Technician A, Technician B Questions

The type of question that is most popularly associated with an ASE test is the "Technician A says . . . Technician B says . . . Who is right?" type. In this type of question, you must identify the correct statement or statements. To answer this type of question correctly, you must carefully read each technician's statement and judge it on its own merit to determine if the statement is true.

Sometimes this type of question begins with a statement about some analysis or repair procedure. This is often referred to as the stem of the question and provides the setup or background information required to understand the conditions the question is based on. This is followed by two statements about the cause of the concern, proper inspection, identification, or repair choices. You are asked whether the first statement, the second statement, both statements, or neither statement is correct.

Analyzing this type of question is a little easier than the other types because there are only two ideas to consider although there are still four choices for an answer.

Technician A, Technician B questions are really double true or false questions. The best way to analyze this kind of question is to consider each technician's statement separately. Ask yourself, is A true or false? Is B true or false? Then select your answer from the four choices. An important point to remember is that an ASE Technician A, Technician B question will never have Technician A and B directly disagreeing with each other. That is why you must evaluate each statement independently.

An example of this type of question and the analysis of it follows.

A vehicle comes into the shop with a gas gauge that will not register above one half full. When the sending unit circuit is disconnected the gauge reads empty and when it is connected to ground the gauge goes to full. Technician A says that the sending unit is shorted to ground. Technician B says the gauge circuit is working and the sending unit is likely the problem. Who is right?

A. A only
B. B only
C. Both A and B
D. Neither A nor B

Analysis:

Reading of the stem of the question sets the conditions of the customer concern and establishes what information is gained from testing. General knowledge of gauge circuits and test procedures are needed to correctly evaluate the technician's conclusions. Note: Avoid being distracted by experience with unusual or problem vehicles that you may have worked on, Other technicians taking the same test do not have that knowledge, so it should not be used as the basis of your answers.

Technician A is wrong because a shorted to ground sending unit would produce a gauge reading equivalent to the test conditions of a grounding the circuit and produce a full reading.
Technician B is correct because the gauge spans when going from an open circuit to a completely
grounded circuit. This would tend to indicate that the problem had to be in the sending unit.
Answer C is not correct. Both technicians are identifying the problem as a sending unit but technician A qualified the problem as a specific type of failure (grounded) that would not have caused the symptoms of the vehicle.
Answer D is not correct because technician B's diagnosis is a possible cause of the conditions identified.

Most-Likely Questions

Most-Likely questions are somewhat difficult because only one choice is correct while the other three choices are nearly correct. An example of a Most-Likely-cause question is as follows:

The Most-Likely cause of reduced turbocharger boost pressure may be a:

A. wastegate valve stuck closed.
B. wastegate valve stuck open.
C. leaking wastegate diaphragm.
D. disconnected wastegate linkage.

Analysis:

Answer A is wrong. A wastegate valve stuck closed increases turbocharger boost pressure.

Answer B is correct. A wastegate valve stuck open decreases turbocharger boost pressure.
Answer C is wrong. A leaking wastegate valve diaphragm increases turbocharger boost pressure.
Answer D is wrong. A disconnected wastegate valve linkage will increase turbocharger boost pressure.

LEAST-Likely Questions

Notice that in Most-Likely questions there is no capitalization. This is not so with LEAST-Likely type questions. For this type of question, look for the choice that would be the LEAST-Likely cause of the described situation. Read the entire question carefully before choosing your answer. An example is as follows:

What is the LEAST-Likely cause of a bent pushrod?

A. Excessive engine speed
B. A sticking valve
C. Excessive valve guide clearance
D. A worn rocker arm stud

Analysis:

Answer A is wrong. Excessive engine speed may cause a bent pushrod.
Answer B is wrong. A sticking valve may cause a bent pushrod.
Answer C is correct. Excessive valve clearance will not generally cause a bent pushrod.
Answer D is wrong. A worn rocker arm stud may cause a bent pushrod.

You should avoid relating questions to those unusual situations that you may have encountered and answer based on the technical and mechanical possibilities.

Summary

There are no four-part multiple-choice ASE questions having "none of the above" or "all of the above" choices. ASE does not use other types of questions, such as fill-in-the-blank, completion, true-false, word-matching, or essay. ASE does not require you to draw diagrams or sketches. If a formula or chart is required to answer a question, it is provided for you. There are no ASE questions that require you to use a pocket calculator.

Overview of the Task List

Exhaust Systems (Test X1)

The Exhaust Systems Test (X1) is not a stand-alone certification. Instead, it serves as a technical complement to certifications in Brakes (A5) and Steering and Suspension (A4). When combined with these certifications, a passing score on the X1 test earns the credential of Undercar Specialist. Certification in A4 and A5 are prerequisites for the X1 exam.

The following section includes the task areas and task lists for this test and a written overview of the topics covered in the test. The task list describes the actual work you should be able to do as a technician that you will be tested on by the ASE. This is your key to the test and you should review this section carefully. We have based our sample test and additional questions upon these tasks, and the overview section will also support your understanding of the task list. ASE advises that the questions on the test may not equal the number of tasks listed; the task lists tell you what ASE expects you to know how to do and be ready to be tested upon.

At the end of each question in the Sample Test and Additional Test Questions sections, a letter and a number will be used as a reference back to this section for additional study. Note the following example: **A.1.3.**

A. Exhaust System Inspection and Repair (11 Questions)

Task A.1 Inspection (6 Questions)

Task A.1.3 Inspect exhaust subsystems [air injection reactor (AIR), exhaust gas recirculation (EGR), oxygen sensor(s) (O_2S/HO_2S), heat riser/early fuel evaporation (EFE), turbochargers] and mounting hardware; determine needed repair.

Example:

1. Which All of the following are typical emission control systems **EXCEPT:**
   A. EGR
   B. PCV
   C. EPA
   D. EFE

(A.1.3)

Analysis:

Question #1
Answer A is wrong. EGR is an emission control for NOx emissions.
Answer B is wrong. PCV is the first emission device and controls combustion gases in the crankcase.
Answer C is correct. Answer C is the Environmental Protection Agency and although EPA has much to do with the reduction of exhaust emissions, it is not an emission control device.
Answer D is wrong. The EFE is for the fuel tank vapor leak feedback to the computer.

Task List and Overview

A. Exhaust System Inspection and Repair (11 Questions)

Task A.1 Inspection (6 Questions)

Task A.1.1 Inspect all exhaust system components for noises, rattles, missing parts, configuration, and routing by visual, audible, and thump testing; determine needed repair.

Major parts of the exhaust system include the exhaust manifold, exhaust pipe, catalytic converters, muffler, and tailpipe. Also, V-type engines have a crossover pipe and an intermediate pipe (connecting the crossover pipe to the muffler). Sometimes a resonator, a secondary silencing device, is installed. All the parts of the system are designed to conform to the available space of the vehicle's undercarriage and yet be a safe distance above the road.

The exhaust system of an engine is designed to conduct the burned gases (exhaust) to the rear of the car and into the air. This system also serves to silence the sounds of combustion. In order to reduce the noise of the combustion of an engine, exhaust gases from the engine are passed through a muffler. The muffler is designed so that the gases expand slowly.

Muffler design must be such that there is the least amount of backpressure developed. Backpressure prevents free flow of the exhaust gases from the engine. As a result, not all of the burned gases will be forced from the cylinders. Remaining exhaust gases dilute the incoming air-fuel mixture and engine power is reduced. The exhaust of a car passes into the exhaust manifold. Then the gas passes through the exhaust pipe into the muffler.

From the muffler it passes into the tailpipe and from there into the atmosphere. Also, some cars use resonators in the system. A certain amount of room for expansion and cooling of the exhaust gas is designed into the exhaust manifold and exhaust pipe. They provide from two to four times the exhaust volume of a single cylinder of the engine. Additional expansion is provided for in the muffler.

Most parts of the exhaust system, particularly the exhaust pipe, muffler, and tailpipe, are subject to rust, corrosion, and cracking. Broken or loose clamps and hangers can allow parts to separate or hit the road as the car moves.

The complete exhaust system should be inspected each time the vehicle is on a lift. Look at all hangers and support points, and check rubber straps and "doughnuts," which help reduce vibration and allow for flexing of the system. Remember, if a hanger breaks, it will add stress and weight to the remaining hangers, which, in turn, may also break.

Any exhaust system inspection should include listening for hissing or rumbling that would result from a leak in the system. An on-lift inspection should pinpoint any of the following types of damage: holes, road damage, separated connections, and bulging muffler seams; kinks and dents; discoloration, rust, and soft corroded metal; torn, broken, or missing parts, hangers, and clamps; and loose tailpipes or other components. A bluish or brownish catalytic converter shell indicates overheating.

Sound out the system by gently tapping the pipes and the muffler with a hammer or mallet. A good part will have a solid metallic sound. A weak or worn-out part will have a dull sound. Listen for falling rust particles on the inside of the muffler. Mufflers usually corrode from the inside out, so it might be worn out on the inside but with the wear not visible from the outside. Remember that some rust spots might be only surface rust.

Grab the tailpipe (when it is cool) and try to move it up and down and side to side. There should be only slight movement in any direction. If the system feels wobbly or loose, check clamps and hangers that fasten the tailpipe to the vehicle. Check all pipes for kinks and dents that might restrict flow of exhaust gases. Take a close look at each connection, including the one between the exhaust manifold and the exhaust pipe.

Task A.1.2 Inspect exhaust system for leaks, restrictions, and overheating by visual, audible, backpressure, vacuum, and temperature testing; determine needed repair.

It is important that no leaks occur in the exhaust system. Exhaust gases contain carbon monoxide (CO). When CO finds its way inside the car, it causes headaches, drowsiness, and nausea. As the amount of CO is increased, unconsciousness and death result. Surveys show that about 5 percent of the cars on the road contain enough carbon monoxide to cause drowsiness and impair driver judgment and reflexes if the windows are rolled up. Any leaks that occur in the exhaust system, from the exhaust manifold to the tailpipe, should be repaired as soon as possible.

Exhaust system components are subject to both physical and chemical damage. Any physical damage to an exhaust system part that causes a partially restricted or blocked exhaust system usually results in loss of power or backfire up through the throttle plate(s). In addition to improper engine operation, a blocked or restricted exhaust system causes increased noise and air pollution. Leaks in the exhaust system caused by either physical or chemical (rust) damage could result in illness, asphyxiation, or even death.

Remember that vehicle exhaust fumes can be very dangerous to one's health. Any perforation of an exhaust system component, of course, requires its replacement. Such leaks not only create the possibility of the vehicle failing emissions testing, but may also create a dangerous condition on the road. Loose joints or fittings should be tightened or corrected with new clamps. Dented pipes, mufflers, and catalytic converters cause restricted exhaust flow and should be replaced if the dents are severe.

Before making a visual inspection, listen closely for hissing or rumbling that indicates beginning of exhaust system failure. With the engine idling, slowly move along the entire system and listen for leaks. To inspect an exhaust system, raise the car on a lift. Using a droplight, closely inspect the system for problems. Inspect all exhaust system joints for leaks. Look for soot and discoloration, which indicate escaping hot gases. These leaks must be corrected to enable the vehicle to pass emissions testing and, more importantly, to protect the vehicle's occupants from being exposed to toxic gases. Pay particular attention to the muffler and all pipe connections, gaskets, and pipe bends. Exhaust leaks will often show up as gray or white carbon lines coming from openings. Catalytic converters can overheat. Look for bluish or brownish discoloration of the outer stainless steel shell. Also, look for blistered or burnt paint or undercoating above and near the converter.

Check for the following:

- Holes and road damage
- Discoloration and rust
- Carbon smudges
- Bulging muffler seams
- Interfering rattle points
- Torn or broken hangers and clamps
- Missing or damaged heat shields

Often leaks and rattles are the only things looked for in an exhaust system. The exhaust system should also be tested for blockage and restrictions. Collapsed pipes or clogged converters and/or mufflers can cause these blockages.

There are many ways to check for a restricted exhaust; the most common of these is the use of a vacuum gauge. Connect a vacuum gauge to an intake manifold vacuum source. Attach vacuum gauge to the intake manifold. Start the engine and observe the vacuum gauge with the engine at idle. It should indicate a vacuum of 16-20 in.Hg. Bring the engine to a moderate speed (about 2,000 rpm) and hold it there. Watch the vacuum gauge. If everything is right, the vacuum reading will be high and will either stay at that reading or increase slightly as the engine runs at this speed. If the vacuum does not build up to at least the idle reading, the exhaust system is restricted or blocked.

The greatest cause of exhaust system failure is from internal corrosion by condensed acid moisture that is chemically produced by the combustion of air and gasoline. Acid moisture condensate is most destructive when both the engine and the exhaust system parts are cold, as in stop and go driving before the engine has a chance to warm up. When the engine and the exhaust system are fully warmed up, the hot gases coming through the system never have a chance to condense on the exhaust system and corrode it.

As you might expect, the parts closest to the engine heat up first and deteriorate more slowly than those farther back in the system. Exhaust pipes tend to have a much longer life expectancy than mufflers or tailpipes that take longer to warm up and collect more of the acid condensate.

Task A.1.3 Inspect exhaust subsystems [air injection reactor (AIR), exhaust gas recirculation (EGR), oxygen sensor(s) (O_2S/HO_2S), heat riser/early fuel evaporation (EFE), turbochargers] and mounting hardware; determine needed repair.

In the past, one of the chief contributors to air pollution was the automobile. For some time now, engines have been engineered to emit very low amounts of certain pollutants. The pollutants that have been drastically reduced are hydrocarbons (HC), carbon monoxide (CO), and oxides of nitrogen (NO_x). HC emissions are caused largely by unburned fuel from the combustion chambers. HC emissions can also originate from evaporative sources such as the gasoline tank. CO emissions are a by-product of the combustion process, resulting from incorrect air-fuel mixtures. NO_x emissions are caused by nitrogen and oxygen uniting at cylinder temperatures above 2,500°F (1,371°C). The Environmental Protection Agency (EPA) establishes emissions standards that limit the amount of these pollutants a vehicle can emit.

To meet these standards, many changes have been made to the engine itself. Plus there have been systems developed and added to the engines to reduce the pollutants they emit. Emission controls on cars and trucks have one purpose: to reduce the amount of pollutants and environmentally damaging substances released by the vehicles. Smog not only appears as dirty air, it is also an irritant to your eyes, nose, and throat. The things necessary to form photochemical smog are HC and NO_x exposed to sunlight in stagnant air. When there is enough HC in the air, it reacts with the NO_x in the air. The energy of sunlight causes these two chemicals to react and form photochemical smog.

The following are those common pollution (emission) control devices that are tied into the exhaust system.

- *Exhaust gas recirculation (EGR)* system. Introduces exhaust gases into the intake air to reduce the temperatures reached during combustion. This reduces the chances of forming NO_x during combustion.
- *Catalytic converter.* Located in the exhaust system, it allows for the burning or converting of HC, CO, and NO_x into harmless substances, such as water.
- *Air-injection system (AIR).* This system reduces HC emissions by introducing fresh air into the exhaust stream to cause minor combustion of the HC in the engine's exhaust.
- *Early fuel evaporation (EFE) system.* This system heats the fuel as it is delivered to a cold engine. This heating of the fuel prevents fuel condensation and reduces HC and CO during cold operation.
- *Oxygen sensors (O_2S).* These sensors are not really emission control devices; rather they play an important role in the regulation of the air-fuel mixture. Precise metering of the fuel going to the cylinders will reduce HC, CO, and NO_x.

Task A.1.4 Visually inspect exhaust system for evidence of tampering (missing/modified and/or improperly installed components); determine needed repair.

It is against the law for anyone to tamper, remove, or intentionally damage any part of the emission control system. When inspecting an exhaust system, carefully look at the system and attempt to identify any signs of tampering, missing/modified, and/or improperly installed components. Sometimes the law violations are obvious, such as the presence of a straight pipe where the catalytic converter should be. Some tampering is not as quickly seen, such as the enlargement of the fuel filler neck so that it can accept leaded fuels. Use of leaded fuels is deadly to the catalytic converter. Also check the piping to the AIR and EGR systems. Any tampering must be noted prior to doing any repair work and must be corrected during the repair procedure. Failure to follow these guidelines would make you liable for any and all fines levied against the owner of the vehicle.

Task A.1.5 Inspect exhaust system electrical components; determine needed repair.

The electrical connections at or in an exhaust system are most likely to be oxygen sensors. Oxygen sensors (O_2S) produce a voltage based on the amount of oxygen in the exhaust. Large amounts of oxygen result from lean mixtures and result in low voltage output from the O_2S. Rich mixtures release lower amounts of oxygen in the exhaust; therefore the O_2S voltage is high. The engine must be at normal operating temperature before the oxygen sensor is tested.

The exhaust gas oxygen sensor is the key sensor in the closed-loop mode. Its input is used by the computer to maintain a balanced air-fuel mixture. The O_2S is threaded into the exhaust manifold or into the exhaust pipe near the engine.

One type of oxygen sensor, made with a zirconium dioxide element, generates a voltage signal proportional to the amount of oxygen in the exhaust gas. It compares the oxygen content in the exhaust gas with the oxygen content of the outside air. As the amount of unburned oxygen in the exhaust gas increases, the voltage output of the sensor drops. Sensor output ranges from 0.1 volt (lean) to 0.9 volt (rich). A perfectly balanced air-fuel mixture of 14.7:1 produces an output centered around 0.5 volt and cycling. When the sensor reading is lean, the computer enriches the air-fuel mixture to the engine. When the sensor reading is rich, the computer leans the air-fuel mixture.

Because the oxygen sensor must be hot to operate, most late-model engines use heated oxygen sensors (HO_2S). These sensors have an internal heating element that allows the sensor to reach operating temperature more quickly and to maintain its temperature during periods of idling or low engine load.

A second type of oxygen sensor, a titania-type, does not generate a voltage signal. Instead it acts like a variable resistor, altering a base voltage supplied by the control module. When the air-fuel mixture is rich, sensor resistance is low. When the mixture is lean, resistance increases. Variable-resistance oxygen sensors do not need an outside air reference. This eliminates the need for internal venting to the outside. They feature very fast warmup, and they operate at lower exhaust temperatures.

The accuracy of the oxygen sensor reading can be affected by air leaks in the intake or exhaust manifold. A misfiring spark plug that allows unburned oxygen to pass into the exhaust also causes the sensor to give a false lean reading.

Before testing an O_2S, refer to the correct wiring diagram to identify the terminals at the sensor. Most late-model engines use heated oxygen sensors (HO_2S). These sensors have an internal heater that helps to stabilize the output signals. Most heated oxygen sensors have four wires connected to them. Two are for the heater and the other two are for the sensor.

An O_2S can be checked with a voltmeter. Connect it between the O_2S wire and ground. The sensor's voltage should be cycling from low voltage to high voltage. The signal from most oxygen sensors varies between 0 and 1 volt. If the voltage is continually high, the air-fuel ratio may be rich or the sensor could have a shorted heater circuit. If the O_2S voltage cycles slowly from rich to lean, a sensor contaminated with RTV sealant, antifreeze, or carbon may be indicated. When the O_2S voltage is continually low, the air-fuel ratio may be lean, the sensor may be defective, or the wire between the sensor and the computer may have a high-resistance problem. If the O_2S voltage signal remains in a mid-range position, the computer may be in open loop or the sensor may be defective.

If a defect in the O_2S signal wire is suspected, backprobe the sensor signal wire at the computer and connect a digital voltmeter from the signal wire to ground with the engine idling. The difference between the voltage readings at the sensor and at the computer should not exceed the vehicle manufacturer's specifications. A typical specification for voltage drop across the average sensor wire is 0.02 volts.

Now check the sensor's ground. With the engine idling, connect the voltmeter from the sensor case to the sensor ground wire on the computer. Typically the maximum allowable voltage drop across the sensor ground circuit is 0.02 volts. Always use the vehicle manufacturer's specifications. If the voltage drop across the sensor ground exceeds specifications, repair the ground wire or the sensor ground in the exhaust manifold.

Most late-model engines are fitted with heated oxygen sensors. If the O_2S heater is not working, the sensor warmup time is extended and the computer stays in open loop longer. In this mode, the computer supplies a richer air-fuel ratio. As a result, the engine's emissions are high and its fuel economy is reduced. To test the heater circuit, disconnect the O_2S connector and connect a voltmeter between the heater voltage supply wire and ground. With the ignition switch on, 12 volts should be supplied on this wire. If the voltage is less than 12 volts, repair the fuse in this voltage supply wire or the wire itself.

With the O_2S wire disconnected, connect an ohmmeter across the heater terminals in the sensor connector. If the heater does not have the specified resistance, replace the sensor.

A faulty O_2S can cause many different types of problems. It can cause excessively high HC and CO emissions and all sorts of drivability problems. Most computer systems monitor the activity of the O_2S and store a code when the sensor's output is not within the desired range. Again, the normal range is between 0 and 1 volt, and the sensor should constantly toggle from close to 0.2 volts to 0.8 volts then back to 0.2. If the range that the sensor toggles in is within the specifications, the computer will think everything is normal and respond accordingly. This, however, does not mean the sensor is working properly.

The activity of an O_2S is best monitored with a labscope. The scope is connected to the sensor in the same way as a voltmeter. The switching of the sensor should be seen as the sensor signal goes to lean to rich to lean continuously.

The voltage signal from an O_2S should have 2 to 3 cross counts with the engine without a load at 2,000 rpm. O_2 signal cross counts are the number of times the O_2 voltage signal changes above or below 0.45 volts in a second. If there are not enough cross counts, the sensor is contaminated or lazy. It should be replaced.

If the sensor's voltage toggles between zero volts and 500 millivolts, it is toggling within its normal range but it is not operating normally. It is biased low or lean. As a result, the computer will be constantly adding fuel to try to reach the upper limit of the sensor. Something is causing the sensor to be biased lean. If the toggling only occurs at the higher limits of the voltage range, the sensor is biased rich. In either case, the computer does not have true control of the air-fuel mixture because of the faulty O_2 signals.

Also keep in mind that on an air pump-equipped car, it is a good idea to disable the air pump before doing this test. Unwanted air may bias the results.

The activity of the sensor can also be monitored on a scanner. By watching the scanner while the engine is running, the O_2 voltage should move to nearly 1 volt then drop back to close to zero volts. Immediately after it drops, the voltage signal should move back up. This immediate cycling is an important function of an O_2S. If the response is slow, the sensor is lazy and should be replaced. With the engine at about 2,500 rpm, the O_2S should cycle from high to low 10 to 40 times in 10 seconds. When testing the O2S, make sure the sensor is heated and the system is in closed loop.

The output from an O_2S should constantly cycle between high and low voltages as the engine is running in closed loop. This cycling is the result of the computer constantly correcting the air-fuel ratio in response to the feedback from the O_2S. When the O_2S reads lean, the computer will richen the mixture. When the O_2S reads rich, the computer will lean the mixture. When the computer does this, it is in control of the air-fuel mixture. Many things can occur to take that control away from the computer. One of these is a faulty O_2S.

If the HO_2S wiring, connector, or terminal is damaged, the entire oxygen sensor assembly should be replaced. Do not attempt to repair the assembly. In order for this sensor to work properly, it must have a clean air reference. The sensor receives this reference from the air that is present around the sensor's signal and heater wires. Any attempt to repair the wires, connectors, or terminals could result in the obstruction of the air reference and degraded oxygen sensor performance.

Task A.1.6 Inspect engine/transmission mount condition and alignment; determine needed repair.

Worn, damaged, or broken engine/transmission mounts could be the cause of exhaust system breakage and rattles. A quick check of the mounts can be done by pulling up and pushing down on the engine or transmission while watching the mount. If the mount's rubber separates from the metal plate or if the engine or transmission case moves up but not down, replace the mount. If there is movement between the metal plate and its attaching point on the frame, tighten the attaching bolts to the appropriate torque and recheck. If it is necessary to replace the mounts, make sure you follow the procedure for maintaining the alignment of the drive line. Failure to do this may result in poor transmission operation and excessive exhaust system rattles. Some manufacturers recommend the use of a special holding fixture or bolt to keep the engine/transmission in place while the mount is being installed.

Task A.2 Repair (5 Questions)

Task A.2.1 Repair or replace failed or damaged mufflers, pipes, and related components.

Sound is vibration in the air. Each of the engine's exhaust valves releases a burst of pressurized exhaust every two turns of the crankshaft. The resulting noise from all of the cylinders blending together results in a loud roar. The roar is really many pressure bursts or vibrations.

The muffler's main task is silencing exhaust gases. It is a combination of tuning chambers, baffle partitions, and ventilated and solid tubes that are designed to absorb noise pulses and move the gases smoothly out the tailpipe. Their location can vary considerably, but most mufflers are located toward the rear of the car.

Specific mufflers are designed for specific cars and engines. Installing a muffler on a car that fits the space, but is not designed for that car or engine, can hurt the car's performance and will not silence as well. It can also cause damage if it develops too much backpressure.

The resonator is a second, usually smaller silencing element in some exhaust systems where the space available under the car does not permit installation of a muffler that can completely silence the exhaust. The resonator smoothes out any loudness or roughness from the first muffler. Most resonators are located to the rear of the muffler.

The tailpipe is the end of the pipeline carrying exhaust fumes to the atmosphere beyond the back end of the car. Heat shields protect vehicle parts from exhaust system heat. They are usually made of pressed or perforated sheet metal. Clamps, brackets, and hangers join and support exhaust system components. The exhaust pipe is a metal pipe–either aluminized steel, stainless steel, or zinc-plated heavy-gauge steel–that runs under the vehicle between the exhaust manifold.

There are usually three exhaust pipes: the header pipe or exhaust pipe, an intermediate pipe between the muffler and catalytic converter, and tailpipe. The exhaust pipe carries collected gases and vapor from the exhaust manifold to the next component in the exhaust system. A "Y" pipe is an exhaust pipe that connects both exhaust manifolds of a V-type engine to form a single exhaust system. An "H" pipe consists of right and left exhaust pipes connected by a balance pipe that forms a dual-exhaust system. Also known as an extension or connecting pipe, the intermediate pipe connects the exhaust pipe with the muffler or resonator. Some converter systems do not have intermediate pipes. The tailpipe carries exhaust gases and vapor out into the air and directs them where they cannot enter the passenger compartment.

Some types of emission control, such as an EGR valve, early fuel evaporation (EFE) valve, or a heat riser valve, are usually mounted between the exhaust manifold and the exhaust pipe. In electronic engine-control systems, an oxygen sensor is installed in the exhaust pipe. This device is used to sense the amount of oxygen in the exhaust and it sends signals to the electronic control unit. This information is used to control the mixture of air and fuel being delivered to the engine.

Because of the rusting that occurs in the exhaust system, parts in the exhaust system often need to be replaced. The parts most often replaced are the muffler and tailpipe. Their life depends on the type of service in which the car is used. If it is used for short trips, it is not uncommon for the muffler and tailpipe to be replaced by 20,000 miles.

Before beginning work on an exhaust system, make sure it is cool to the touch. Some technicians disconnect the battery's negative cable before starting to work to avoid short-circuiting the electrical system. Soak all rusted bolts, nuts, etc., with a good penetrating oil. Finally, check the system for critical clearance points so they can be maintained when new components are installed.

Most exhaust work involves the replacement of parts. When replacing exhaust parts, make sure the new parts are exact replacements for the original parts. Doing this will ensure proper fit and alignment, as well as provide acceptable noise levels.

While trying to replace a part in the exhaust system, you may run into parts that are rusted together. This is especially a problem when a pipe slips into another pipe of muffler. If you are trying to reuse one of the parts, you should carefully use a cold chisel or slitting tool on the outer pipe of the rusted union. You must be careful when doing this because you can easily damage the inner pipe. It must be perfectly round to seal in a new pipe.

Slide the new pipe over the old. Position the rest of the exhaust system so that all clearances are evident and the parts aligned, then put a U-clamp over the new outer pipe to secure the connection.

To replace a damaged exhaust pipe, begin by supporting the converter to keep it from falling. Carefully remove the oxygen sensor if there is one. Remove any hangers or clamps holding the exhaust pipe to the frame. Unbolt the flange holding the exhaust pipe to the exhaust manifold. When removing the exhaust pipe, check to see if there is a gasket. If so, discard it and replace it with a new one. Once the joint has been taken apart, the gasket loses its effectiveness. Disconnect the pipe from the converter and pull the front exhaust pipe loose and remove it.

Although most exhaust systems use flanges or a slip joint and clamps to fasten the pipe to the muffler, a few use a welded connection. If the vehicle's system is welded, cut the pipe at the joint with a hacksaw or pipe cutter. The new pipe need not be welded to the muffler. An adapter, available with the pipe, can be used instead. When measuring the length for the new pipe, allow at least 2 inches (50.8 mm) for the adapter to enter the muffler.

The joint between the manifold and exhaust pipe is of the flange-and-gasket type. Brass nuts are used to hold the flanges together. Brass nuts are used, as they will not rust to the stud, which makes removal easy. Exhaust system parts can be obtained from a supplier or by keeping a supply of parts in the shop. However, stocking enough exhaust pipes, mufflers, and tailpipes takes much space and involves a large amount of money. To help overcome these problems, tube-bending equipment is available. With this equipment, only straight tubing in various diameters is stocked. The straight pieces are then bent to the desired shape.

Many tube-bending tools are fully automatic and are designed to produce bends through 3 inches outside diameter tubing. A heavy-duty expander is used on the ends of the cut to form slip connections.

The exhaust pipe is designed with the lower end slightly larger in diameter than the opening in the muffler. The muffler opening can then be slipped into the end of the exhaust pipe. A clamp is placed around the end of the exhaust pipe. When tightened, the two parts are held together. Metal or metal-and-fabric straps are used to hold the muffler and pipes in proper alignment.

To replace a muffler and tailpipe, remove the clamps and straps holding the tailpipe in place. Next, remove the tailpipe from the muffler. The tailpipe rusts to the muffler, which makes it hard to separate at the joint. If any of the parts are to be used again, penetrating oil should be applied to the joints before pulling them apart. In most cases, exhaust systems are not used again, and, therefore, can be cut apart.

Hacksaws can be used for muffler removal, but power-driven tools will do the job much more quickly. It may be necessary to expand the end of a muffler pipe, tailpipe, or exhaust pipe. This is so the pipes can be assembled.

When repairing an exhaust system, remember the following:

- Use rust penetrant on all threaded fasteners that will be reused. This is especially important on the exhaust manifold flange nuts or bolts.
- Use an air chisel, cut-off tool, cutting torch, or hacksaw to remove faulty parts. Make sure you do not damage parts that will be reused.
- A six-point socket and ratchet or an impact wrench will usually allow quick fastener removal without rounding off the fastener heads.
- Wear safety glasses or goggles to keep rust and dirt from entering your eyes.
- Obtain the correct replacement parts.
- A pipe expander should be used to enlarge pipe ends as needed. A pipe shaper can be used to straighten dented pipe ends.
- Position all clamps properly.
- Install any necessary adapters.
- Make sure all pipes are fully inserted.

- Double-check the routing of the exhaust system. Keep adequate clearance between the exhaust system components and the vehicle's body and chassis.
- Tighten all clamps and hangers evenly. Torque the fasteners only enough to hold the parts. Overtightening will smash and deform the pipes, possibly causing leakage.
- When replacing an exhaust manifold, use a gasket and check sealing surface flatness. If the manifold is warped, it must be machined flat. Torque the exhaust manifold bolts to specification.
- Always use new gaskets and O-rings.
- Check heat valve operation using the information in a service manual.
- Install all heat shields.
- Check the system for leaks and rattles after repairs.

Many new vehicle exhaust systems are made of stainless steel. Stainless steel will not rust and will provide much longer service life than ordinary steel systems. When servicing stainless steel exhaust systems, use heavy-duty clamps designed for this type of system. Conventional muffler clamps are not strong enough to make a good connection.

When cutting or welding stainless steel, use the correct rod or wire material. Keep in mind that stainless steel does not react in the same manner as carbon steel when heated near its melting point. Stainless steel can be "red hot" when it looks cool.

Task A.2.2 Repair or replace damaged catalytic converters.

A catalytic converter contains a ceramic element coated with a catalyst. A catalyst is something that causes a chemical reaction without being part of the reaction. A catalytic converter causes a chemical change to take place in the passing exhaust gases. Most of the harmful gases are changed to harmless gases.

Three different materials are used as the catalyst in automotive converters: platinum, palladium, and rhodium. Platinum and palladium are the oxidizing elements of a converter. When HC and CO are exposed to heated surfaces covered with platinum and palladium, a chemical reaction takes place. The HC and CO are combined with oxygen to become H_2O and CO_2. Rhodium is a reducing catalyst. When NO_x is exposed to hot rhodium, oxygen is removed and NO_x becomes just N. The removal of oxygen is called reduction, which is why rhodium is a reducing catalyst.

A catalytic converter that contains all three catalysts and reduces HC, CO, and NO_x is called a three-way converter. Catalytic converters that affect only HC and CO are called oxidizing converters. Three-way converters have the oxidizing catalysts in part of the container and the reducing catalyst in the other. Fresh air is injected by the secondary air system between the two catalysts. This air helps the oxidizing catalyst work by making extra oxygen available. The air from the secondary air system is not always forced into the converter. Rather it is controlled by the secondary air system. Fresh air added to the exhaust at the wrong time could produce NO_x, something the catalytic converter is trying to destroy.

There are several types of catalytic converters. The most common follow.

- Pellet Catalytic Converter–Uses hundreds of small beads that act as the catalyst agent
- Monolithic Catalytic Converter–Uses a ceramic (made of substrate material called alumina) block shaped like a honeycomb that is coated with a special chemical that helps the converter act on the exhaust gases
- Three-way Converter–The dual-bed type treats all three controlled emission gases. It oxidizes HC and CO by adding oxygen and reduces NO_x by removing oxygen from the nitrogen oxides.
- Mini-catalytic Converter–It provides a close coupled converter that is either built in the exhaust manifold or located next to it. It is primarily used to clean the exhaust during engine warmup. These converters are commonly called warmup converters or preheaters.
- Particulate Oxidizer Catalytic Converter–Uses a monolithic element, located between the exhaust manifold and turbocharger, to react to the particulates of a diesel engine Particulates are solid particles of carbon-like soot that are emitted from a diesel engine as black smoke.
- Many catalytic converters have an air hose connected from the AIR system to the oxidizing catalyst. This air helps the converter work by making extra oxygen available.

The air from the AIR system is not always forced into the converter; rather it is controlled by the vehicle's PCM. Fresh air added to the exhaust at the wrong time could overheat the converter and produce NO_x, something the converter is trying to destroy.

OBD-II regulations call for a way to inform the driver that the vehicle's converter has a problem and may be ineffective. The PCM monitors the activity of the converter by comparing the signals of an HO_2S located at the front of the converter with the signals from a HO_2S located at the rear. If the sensors' outputs are the same, the converter is not working properly and the malfunction indicator lamp (MIL) on the dash will light.

The converter is normally a trouble-free emission control device, but two things can damage it. One is leaded gasoline. Lead coats the catalyst and renders it useless. The difficulty of obtaining leaded gasoline has reduced this problem. The other is overheating. If raw fuel enters the exhaust because of a fouled spark plug or other problem, the temperature of the converter quickly increases. The heat can melt the ceramic honeycomb or pellets inside, causing a major restriction to the flow of exhaust.

If a catalytic converter is found to be bad, it is replaced. There are two types of replacement. Installation kits include all necessary components for the installation.

There are also direct-fit catalytic converters. Under no circumstances should the converter be removed and replaced with a straight piece of pipe (a test pipe). This practice is illegal for the professional auto technician. Because of constant change in EPA catalytic converter removal and installation requirements, check with the manufacturer or EPA for the latest data regarding replacement.

If you are replacing a plugged converter, it is important to remember it failed because there is another problem that needs to be diagnosed. Replacing the converter will eliminate the restriction and restore proper emissions performance. However, the new converter will likely suffer the same fate as the old one unless you figure out what caused the old converter to get too hot. Look for things like fouled spark plugs, bad plug wires, a cylinder with low or no compression, or a computerized feedback system that stays in open loop all the time (usually a bad coolant sensor or O_2 sensor).

On dual-converter systems that have a converter for each side of a V8 engine, the side with the bad converter will tell you which cylinder bank to check. If the converter on the right is bad, for example, check the O_2 sensor, spark plugs, and compression on the right cylinder bank.

Another item that should be replaced is the oxygen sensor. To do its job efficiently, the converter needs an air-fuel mixture that is constantly flip-flopping back and forth from rich to lean. If the O_2 sensor is sluggish or dead, the fuel feedback loop will either flip-flop too slowly or remain rich all the time. Either way it will increase CO emissions and converter temperature.

A bad coolant sensor that always indicates a cold engine (or an open coolant sensor circuit) will also keep the fuel system in open loop, which means a steady diet of excess fuel and poor converter performance. A thermostat that is stuck open or is too cold for the application can also cause the same sort of problem.

It is also important to check out the air pump (or aspirator valve) and related plumbing because these components provide fresh air for the converter to reburn the pollutants in the exhaust. If the air pump is not working, or the air is not getting to the converter in the right amounts at the right time because of a bad diverter valve or damaged or leaky plumbing, it can reduce the converter's operating efficiency significantly.

Task A.2.3 Repair or replace exhaust manifolds.

The exhaust manifold is a bank of pipes which collects the exhaust gases from each cylinder and carries them into the rest of the exhaust system. In late-model cars, the exhaust manifold also contains a catalytic converter. In older models, the sole converter was located under the floor between the exhaust pipe and the muffler. In-line engines have one exhaust manifold. V-type engines have an exhaust manifold on each side of the engine. The exhaust pipe runs between the exhaust manifold and the catalytic converter. An exhaust manifold will have three, four, or six passages, depending on the type of engine. These passages blend into a single passage at the other end that connects to an exhaust pipe. From that point, the flow of exhaust gases continues to the catalytic converter, muffler, and tailpipe, and then exits at the rear of the car.

An exhaust manifold bolts to the cylinder head to enclose the exhaust port openings. The manifold is usually made of cast iron. High-flow, high-performance manifolds, which are commonly called headers, are sometimes made of stainless steel or lightweight steel tubing. Exhaust manifolds for most vehicles are made of cast- or nodular-iron.

Some newer vehicles have stamped, heavy-gauge sheet metal or stainless steel units. The cylinder head mating surface is machined smooth and flat. An exhaust manifold gasket is commonly used between the cylinder head and manifold to help prevent leakage. The outlet end of the exhaust manifold has a round opening with holes for stud bolts or cap screws. A gasket or an O-ring (exhaust manifold doughnut) seals the connection between the exhaust manifold outlet and header pipe to prevent leakage.

Exhaust systems are designed for particular engine-chassis combinations. Exhaust system length, pipe size, and silencer size are used to tune the flow of gases within the exhaust system. Proper tuning of the exhaust manifold tubes can actually create a partial vacuum that helps draw exhaust gases out of the cylinder, improving volumetric efficiency. Separate, tuned exhaust headers can also improve efficiency by preventing the exhaust flow of one cylinder from interfering with the exhaust flow of another cylinder.

Cylinders next to one another may release exhaust gas at about the same time. When this happens, the pressure of the exhaust gas from one cylinder can interfere with the flow from the other cylinder. With separate headers, the cylinders are isolated from one another, interference is eliminated, and the engine breathes better. The problem of interference is especially common with V8 engines. However, exhaust headers tend to improve the performance of all engines.

Exhaust manifolds may also be the attaching point for the air-injection reaction (AIR) pipe. This pipe introduces cool air from the AIR system into the exhaust stream. Some exhaust manifolds have provisions for the EGR pipe. This pipe takes a sample of the exhaust gases and delivers it to the EGR valve. Also, some exhaust manifolds have a tapped bore that retains the oxygen sensor.

Before beginning work on an exhaust system, make sure it is cool to the touch. Some technicians disconnect the battery's negative cable before starting to work to avoid short-circuiting the electrical system. Soak all rusted bolts, nuts, etc., with a good penetrating oil. Finally, check the system for critical clearance points so they can be maintained when new components are installed.

Most exhaust system servicing involves the replacement of parts. When replacing exhaust system components, it is important that original equipment parts (or their equivalent) are used to ensure proper alignment with other parts in the system and provide acceptable exhaust noise levels. When replacing only one component in an exhaust system, it is not always necessary to take off the parts behind it.

The manifold itself rarely causes any problems. On occasion, an exhaust manifold will warp because of excess heat. A straightedge and feeler gauge can be used to check the machined surface of the manifold. Another problem that can also result from the high temperatures generated by the engine is a cracked manifold. This usually occurs after the car passes through a large puddle and cold water splashes on the manifold's hot surface.

If the manifold is warped beyond manufacturer's specifications or is cracked, it must be replaced. Also, check the exhaust pipe for signs of collapse. If there is damage, repair it.

These repairs should be done as directed in the vehicle's service manual. If a heated oxygen sensor (HO_2S) is mounted in the exhaust manifold, a cracked manifold upstream from the HO_2S allows air to enter the exhaust manifold. This air affects the HO_2S sensor signal and causes the powertrain control module (PCM) to supply a rich air-fuel ratio.

The exhaust manifold gasket seals the joint between the head and exhaust manifold. Many new engines are assembled without exhaust manifold gaskets. This is possible because new manifolds are flat and fit tightly against the head without leaks. Exhaust manifolds go through many heating/cooling cycles. This causes stress and some corrosion in the exhaust manifold. Removing the manifold will usually distort the manifold slightly so it is no longer flat enough to seal without a gasket. Exhaust manifold gaskets are normally used to eliminate leaks when exhaust manifolds are reinstalled.

The most likely spots for leaking gaskets and seals are between the exhaust manifold and the cylinder head and between the exhaust pipe and the exhaust manifold. When installing exhaust gaskets, carefully follow the recommendations on the gasket package label and instruction forms. Read through all installation steps before beginning. Take note of any of the original equipment manufacturer's recommendations in service manuals that could affect engine sealing. This is especially important when working with aluminum components. Manifolds warp more easily if an attempt is made to remove them while still hot. Remember heat expands metal, making assembly bolts more difficult to remove and easier to break.

Follow the torque sequence in reverse to loosen each bolt. Then repeat the process again to remove the bolts. This minimizes the chance of components warping.

Any debris left on the sealing surfaces increases the chance of leaks. A good gasket remover will quickly soften the old gasket debris and adhesive for quick removal. Carefully remove the softened pieces with a scraper and a wire brush. Be sure to use a nonmetallic scraper when attempting to remove gasket material from aluminum surfaces.

Inspect the manifold for irregularities that might cause leaks, such as gouges, scratches, or cracks. Replace any parts that are cracked or badly warped. This will insure proper sealing of the manifold.

Due to high heat conditions, it is important to re-tap and re-die all threaded bolt holes, studs, and mounting bolts. This procedure insures tight, balanced clamping forces on the gasket. Lubricate the threads with a good high-temperature anti-seize lubricant. Use a small amount of contact adhesive to hold the gasket in place. Align the gasket properly before the adhesive dries. Allow the adhesive to dry completely before proceeding with manifold installation.

Install the bolts finger tight. Tighten the bolts in three steps–one-half, three-quarters, and full torque–following the torque tables in the service manual or gasket manufacturer's instructions. Torquing is usually begun in the center of the manifold, working outward in an X pattern.

Task A.2.4 Repair or replace exhaust system mounting hardware and related installation components.

Clamps, brackets, and hangers are used to properly join and support the various parts of the exhaust system. These parts also help to isolate exhaust noise by preventing its transfer through the frame or body to the passenger compartment. Clamps help to secure exhaust system parts to one another. The pipes are formed in such a way that one slips inside the other. This design makes a close fit. A U-type clamp usually holds this connection tight. Another important job of clamps and brackets is to hold pipes to the bottom of the vehicle. Clamps and brackets must be designed to allow the exhaust system to vibrate without transferring the vibrations through the car. U-bolts of the same size as the exhaust pipe's outside diameter are used to clamp together sections of pipe. One end of one of the pipes is expanded to fit over the other piece of pipe. Clamps provide leak-free connections at joints in the muffler system.

Gaskets, usually made from asbestos, pressed steel, or sintered iron are used to ensure tighter connections between the exhaust manifold and the exhaust pipe. Some exhaust systems are a single unit in which the pieces are welded together by the factory. By welding instead of clamping the assembly together, car makers save the weight of overlapping joints as well as that of clamps.

There are many different types of hangers available. Each is designed for a particular application. Some exhaust systems are supported by doughnut-shaped rubber rings between hooks on the exhaust component and on the frame or car body. Others are supported at the exhaust pipe and tailpipe connections by a combination of metal and reinforced fabric hanger. Both the doughnuts and the reinforced fabric allow the exhaust system to vibrate without breakage that could be caused by direct physical connection to the vehicle's frame.

Hangers hold the clamps, pipes, and muffler to the underside of the car. They are flexible to absorb road vibration, though if they are too flexible, road vibration can break them. On the other hand, if they are too firm, vibration and noise from the exhaust system are transferred to the underbody of the car and can be heard in the passenger compartment.

A hanger is usually a piece of fabric and rubber that resembles a tire. Each end is riveted to a piece of metal that is fastened on one end to the car body or frame. The other end is clamped to the pipe or muffler. The hanger allows the exhaust to be positioned away from other parts of the car and allows some flexibility from engine torque and vibration. Using a piece of rubber isolates the exhaust noise from the car body.

Heat shields are located in areas where the exhaust system components (especially the catalytic converter and muffler) are close to the car body or near the ground. The shields reduce the amount of heat transferred into the car's body and protect items under the vehicle. They are usually made of pressed or perforated sheet metal. Heat shields trap the heat in the exhaust system, which has a direct effect on maintaining exhaust gas velocity.

Always reinstall all exhaust system heat shields. If the heat shields are not installed, undercoating, carpeting, dry leaves on the ground, and other flammable materials could catch on fire!

Care must be used so that there are no kinks or flattened areas in the exhaust system that would obstruct the free flow of the exhaust gases. Any obstruction caused by internal blockage will reduce power and fuel economy. This could be caused by a defective EGR valve/system, clogged catalytic converter, collapsed muffler, frozen heat control valve, and kinks in exhaust pipes.

Task A.2.5 Repair or replace exhaust subsystems [air injection reactor (AIR), exhaust gas recirculation (EGR), oxygen sensor ($0_2S/H0_2S$), heat riser/early fuel evaporation (EFE), turbochargers] and mounting hardware.

Not all engines are equipped with an air-injection system; only those that need them to meet emissions standards have them. Therefore, air-injection systems are vital to proper emissions on engines equipped with them. Each system has its own test procedure; always follow the manufacturer's recommendations for testing. There are three basic designs of air-injection systems.

Many air pumps are driven by a drive belt; others use an electric powered pump. When installing a driven pump, make sure the pulley is aligned with the drive pulley on the engine's crankshaft. The drive belt tension needs to be set to specifications. For both electric and pulley-driven pumps it is extremely important that the mounting bolts are tightened to specifications. When replacing the tubing and/or check and diverter valves, make sure they are tightened properly. Follow the manufacturer's recommendations before using any sealing compound on the threads of any part.

Exhaust gas recirculating (EGR) valves must be sealed to the intake manifold or mounting point. Make sure the mounting surface is clean and true. Always use a new gasket and carefully tighten the retaining bolts. It is best to run the mounting bolts down on each side of the valve before torquing them. Make sure all vacuum lines/hoses and/or electrical connections are made after the valve is installed.

If the O_2S wiring, connector, or terminal is damaged, the entire oxygen sensor assembly should be replaced. Do not attempt to repair the assembly. Oxygen sensors are threaded into the exhaust system. Make sure the threads are not crossed when installing the sensor. Never use a sealing compound on the threads unless advised to do this by the manufacturer. Make sure the sensor's or harness wires are not bent sharply or kinked. Hard bends and kinks could block the reference air path through the lead wire. Also, to prevent damage due to water intrusion, be sure the peripheral seal remains intact on the vehicle harness.

The heat riser valve provides extra heat under the carburetor as the engine is warming up to improve performance. It is located at the outlet on one side of the exhaust manifold, usually on the right side. Governed by a bimetal spring, the valve is closed when the engine is cold. This forces the hot exhaust gases from the right bank of cylinders up through a heat passage located directly under the carburetor. This extra heat under the carburetor improves performance during the time the engine is warming up.

Then as the engine temperature increases, the valve opens and the exhaust gases are discharged from the exhaust manifold into the exhaust system.

A stuck heat riser valve can cause valve burning and carburetor vapor lock. Stuck valves can generally be freed with work and penetrating oil. The valves can also be lubricated with graphite to prevent future seizure. When replacing the heat riser valve, make sure it operates freely after installation and connect all hoses that were disconnected during replacement of the valve.

Turbochargers are replaced as a unit. There are several important things that need to be done during and after installation of a new or rebuilt unit. Make sure the new turbocharger is the correct type. Compare part numbers. Always use new gaskets and seals. Torque all fasteners to specifications.

After replacement of a turbocharger, or after an engine has been unused or stored, there can be a considerable lag after engine startup before the oil pressure is sufficient to deliver oil to the turbocharger's bearings. To prevent this problem, follow these simple steps: (1) Make certain that the oil inlet and drain lines are clean before connecting them. (2) Be sure the engine oil is clean and at the proper level. (3) Fill the oil filter with clean oil. (4) Leave the oil drain line disconnected at the turbocharger, and crank the engine without starting it until oil flows out of the turbocharger drain port. (5) Connect the drain line, start the engine, and operate it at low idle for a few minutes before running it at higher speeds.

B. Emission Systems Diagnosis (8 Questions)

Task B.1 Identify failed catalytic converter(s); determine cause of failure; determine needed repair.

A catalytic converter can present three different problems. It can leak, cause an exhaust restriction, or just not work. In all cases, a defective converter must be replaced. A plugged converter or any exhaust restriction can cause loss of power at high speeds, stalling after starting (if totally blocked), a drop in engine vacuum as engine rpm increases, or sometimes popping or backfiring at the carburetor. If the vehicle has a dual exhaust system with two separate converters, it is possible that only one side will be blocked. The engine will still run but not very well because of the blockage on one side.

If there is more than one converter in a single exhaust path, the front converter will usually be the one that fails first if unburned fuel or contaminants find their way into the exhaust system.

This type of converter failure normally results from ignition misfire or a leaky exhaust valve. Both result in excessive HC in the exhaust. When the fuel hits the converter, it ignites and causes the operating temperature of the converter to drastically increase. The extreme heat melts the material inside the converter causing a partial or complete blockage. Replacing a plugged converter will fix the immediate problem but will not correct the cause of the blockage. Always determine the cause of the converter problem and fix that after replacing the converter.

To verify that the exhaust system or converter is restricting exhaust flow, connect a vacuum gauge to an intake vacuum source. The vacuum is observed when the engine is at fast idle. If the vacuum reading decreases over time, the exhaust is restricted. Another way to check for a restricted exhaust or catalyst is to insert a pressure gauge in the exhaust manifold's bore for the O_2S. With the gauge in place, hold the engine's speed at 2,000 rpm and watch the gauge. The desired pressure reading will be less than 1.25 psi. A very bad restriction will give a reading of over 2.75 psi. But this will only tell you whether there is too much backpressure in the exhaust. It will not tell you where. Keep in mind that exhaust restrictions can also be caused by damaged or crushed pipes or collapsed baffles inside a muffler or resonator. If you see no obvious damage to the exhaust system, disconnect the head pipe, or Y-pipe, where it enters the converter and look inside. Any damage that may have occurred to the honeycomb inside will usually be obvious.

Another way to measure backpressure is to measure it at the air pump check valve by removing the check valve and installing a pressure gauge. The check valve must connect to the exhaust system ahead of the converter; however, if the plumbing hooks up at the converter, this technique will not give you reliable results. Backpressure readings should generally be less than 1.5 psi (though some do allow as much as 2.75 pounds at idle). Rev the engine to 2,000 rpm and note how much the reading increases. If it is higher than 3 psi (or keeps climbing) you have identified a restriction problem.

There are many ways to test a catalytic converter; one of these is to simply smack the converter with a rubber mallet. If the converter rattles, it needs to be replaced and there is no need to do other testing. A rattle indicates loose catalyst substrate, which will soon rattle into small pieces. This is one test and is not used to determine if the catalyst is good.

The converter should be checked for its ability to convert CO and HC into CO_2 and water. There are three separate tests for doing this. The first method is the delta temperature test. To conduct this test, use a hand-held digital pyrometer. By touching the pyrometer probe to the exhaust pipe just ahead of and just behind the converter, there should be an increase of at least 100°F or 8 percent above the inlet temperature reading as the exhaust gases pass through the converter. If the outlet temperature is the same or lower, nothing is happening inside the converter. To do its job efficiently, the converter needs a steady supply of oxygen from the air pump. A bad pump, faulty diverter valve or control valve, leaky air connections, or faulty computer control over the air-injection system could be preventing the needed oxygen from reaching the converter. If the converter fails this test, check those systems.

The next test is called the O_2 storage test and is based on the fact that a good converter stores oxygen. Begin by disabling the air-injection system. Once the analyzer and converter are warmed up, hold the engine at 2,000 rpm. Watch the readings on the exhaust analyzer. Once the numbers stop dropping, check the oxygen level on the gas analyzer. The O_2 readings should be about 0.5 to 1 percent. This shows the converter is using most of the available oxygen. It is important to observe the O_2 reading as soon as the CO begins to drop. If the converter fails the tests, chances are that it is working poorly or not at all.

This final converter test uses a principal that checks the converter's efficiency. Before beginning this test, make sure the converter is warmed up. Calibrate a four- or five-gas analyzer and insert its probe into the tail pipe. Disable the ignition. Then crank the engine for 9 seconds while pumping the throttle. Watch the readings on the analyzer; the CO_2 on fuel-injected vehicles should be over 11 percent and carbureted vehicles should have a reading of over 10 percent. As soon as you get your readings, reconnect the ignition and start the engine. Do this as quickly as possible to cool off the catalytic converter. If, while the engine is cranking, the HC goes above 1,500 ppm, stop cranking; the converter is not working. Also stop cranking once the CO_2 readings reach 10 or 11 percent; the converter is good. If the catalytic converter is bad, there will be high HC and, of course, low CO_2 at the tailpipe. Do not repeat this test more than ONE time without running the engine in between. If a catalytic converter is found to be bad, it is replaced.

OBD-II regulations call for a way to inform the driver that the vehicle's converter has a problem and may be ineffective. The PCM monitors the activity of the converter by comparing the signals of an HO_2S located at the front of the converter with the signals from an HO_2S located at the rear. If the sensors' outputs are the same, the converter is not working properly and the malfunction indicator lamp (MIL) on the dash will light.

OBD-II vehicles use a minimum of two oxygen sensors. One of these is used for feedback to the PCM for fuel control and the other, located at the rear of the catalytic converter, gives an indication of the efficiency of the converter. The downstream O_2S is sometimes called the "catalyst monitor sensor" (CMS).

If the converter is operating properly, the signal from the pre-catalyst O_2S will have oscillations while the post-catalyst O_2S will be relatively flat. Once the signal from the rear sensor approaches that of the front sensor, the MIL comes on and a DTC is set.

The downstream HO_2S sensors have additional protection to prevent the collection of condensation on the ceramic. The internal heater is not turned on until the ECT sensor signal indicates a warmed-up engine. This action prevents cracking of the ceramic. Gold-plated pins and sockets are used in the HO_2S sensors, and the downstream and upstream sensors have different wiring harness connectors.

When the catalytic converter is storing oxygen properly, the downstream HO_2S sensors provide low-frequency voltage signals. If the catalytic converter is not storing oxygen properly, the voltage signal frequency increases on the downstream HO_2S sensors until the frequency of the downstream HO_2S sensors approaches the frequency of the upstream HO_2S sensors. When the downstream HO2S voltage signals reach a certain frequency, a DTC is set in the PCM memory. If the fault occurs on three drive cycles, the MIL light is illuminated.

An OBD-II system continuously monitors the entire emissions system, switches on a MIL if something goes wrong, and stores a fault code in the PCM when it detects a problem. The codes are well defined and can lead a technician to the problem. A scan tool must be used to access and interpret emission-related DTCs regardless of the make and model of the vehicle.

If a misfire that threatens engine or catalyst damage occurs, the misfire monitor flashes the MIL on the first occurrence of the misfire. A fault detected by the catalyst monitor must occur on three drive cycles before the MIL is illuminated. When the fault is no longer present and the MIL is turned off, the DTC is erased after 40 engine warmup cycles. A technician may use a scan tester to erase DTCs immediately. Possible causes for OBD-II catalyst monitor failure are fuel contaminants, leaking exhaust, engine mechanical problems, defective upstream or downstream oxygen sensor circuits, and defective PCM.

Task B.2 Identify failed air injection reactor (AIR) system; determine root cause of failure; determine needed repair.

In some AIR systems, pressure relief valves are mounted in the AIRB and AIRD valves. Other AIR systems have a pressure relief valve in the pump. If the pressure relief valve is stuck open, airflow from the pump is continually exhausted through this valve, which causes high tailpipe emissions.

Some AIR systems will set DTCs in the PCM if there is a fault in the AIRB or AIRD solenoids and related wiring. In some AIR systems, DTCs are set in the PCM memory if the airflow from the pump is continually upstream or downstream. Always use a scan tool to check for any DTCs related to the AIR system, and correct the causes of these codes before proceeding with further system diagnosis.

If the AIR system does not pump air into the exhaust ports during engine warmup, HC emissions are high during this mode, and the O_2S, or sensors, takes longer to reach normal operating temperature. Under this condition, the PCM remains in open loop longer. Since the air-fuel ratio is richer in open loop, fuel economy is reduced.

When the AIR system pumps air into the exhaust ports with the engine at normal operating temperature, the additional air in the exhaust stream causes lean signals from the O_2S, or sensors. The PCM responds to these lean signals by providing a rich air-fuel ratio from the injectors. This action increases fuel consumption. A vehicle can definitely fail an emission test because of air flowing past the O_2S when it should not be. If the O_2S is always sending a lean signal back to the computer, check the air-injection system.

When the engine is started, listen for air being exhausted from the AIRB valve for a short period of time. If this air is not exhausted, remove the vacuum hose from the AIRB and start the engine. If air is now exhausted from the AIRB valve, check the AIRB solenoid and connecting wires. When air is still not exhausted from the AIRB valve, check the air supply from the pump to the valve. If the air supply is available, replace the AIRB valve.

During engine warmup, remove the hose from the AIRD valve to the exhaust ports and check for airflow from this hose. If airflow is present, the system is operating normally in this mode. When air is not flowing from this hose, remove the vacuum hose from the AIRD valve and connect a vacuum gauge to this hose. If vacuum is above 12-in. Hg, replace the AIRD valve. When the vacuum is zero, check vacuum hoses, the AIRD solenoid, and connecting wires.

With the engine at normal operating temperature, disconnect the air hose between the AIRD valve and the catalytic converters and check for airflow from this hose. When airflow is present, system operation in the downstream mode is normal. If there is no airflow from this hose, disconnect the vacuum hose from the AIRD valve and connect a vacuum gauge to the hose. When the vacuum gauge indicates zero vacuum, replace the AIRD valve. If some vacuum is indicated on the gauge, check the hose, the AIRD solenoid, and connecting wires.

Run the engine at idle with the secondary air system on (enabled). Using an exhaust gas analyzer, measure and record the oxygen (O_2) levels. Next, disable the secondary air system and continue to allow the engine to idle. Again, measure and record the oxygen level in the exhaust gases. The secondary air system should be supplying 2 percent to 5 percent more oxygen when it is operational (enabled).

Computer systems with OBD-II will perform certain tests on various subsystems of the engine management system. OBD-II is designed to turn on the MIL when the vehicle has any failure that could potentially cause the vehicle to exceed its designed emission standard by a factor of 1.5. The system does that by the use of a monitor. If one or more monitored systems are out limit, then the MIL turns on to indicate a problem. OBD-II systems have a monitor for checking AIR system operation. The AIR system operation can be verified by turning the AIR system on to inject air upstream of the oxygen sensor while monitoring its signal. Many designs inject air into the exhaust manifold when the engine is in open loop and switch the air to the converter when it is in closed loop. If the air is diverted to the exhaust manifold during closed loop, the O_2S thinks the mixture is lean and the signal should drop.

On some vehicles, the AIR system is monitored with passive and active tests. During the passive test, the voltage of the pre-catalyst HO_2S is monitored from startup to closed-loop operation. The AIR pump is normally on during this time. Once the HO_2S is warm enough to produce a voltage signal, the voltage should be low if the AIR pump is delivering air to the exhaust manifold. The secondary AIR monitor will indicate a pass if the HO_2S voltage is low at this time. The passive test also looks for a higher HO_2S voltage when the airflow to the exhaust manifold is turned off by the PCM. When the AIR system passes the passive test, no further testing is done. If the AIR system fails the passive test or if the test is inconclusive, the AIR monitor in the PCM proceeds with the active test.

During the active test, the PCM cycles the airflow to the exhaust manifold on and off during closed-loop operation and monitors the precatalyst HO_2S voltage and the short-term fuel trim value. When the airflow to the exhaust manifold is turned on, the sensor's voltage should decrease and the short-term fuel trim should indicate a richer condition. The secondary AIR system monitor illuminates the MIL and stores a DTC in the PCM's memory if the AIR system fails the active test on two consecutive trips.

The primary purpose of OBD-II is to make the diagnosis of emissions and drivability problems simple and uniform in the future. No longer is it necessary to learn entirely new systems from each manufacturer. The basic plan is to allow technicians to diagnose any vehicle with the same diagnostic tools. Manufacturers can introduce special diagnostic tools or capacities for their own systems, providing that standard scan tools, along with DMMs and labscopes, can analyze the system. These special tools can have additional capabilities beyond those given in the standards. One of the mandated capabilities is the "freeze frame" or snapshot feature. This is the ability of the system to record data from all of its sensors and actuators at a time when the system turns on the MIL.

The MIL is on the instrument panel. The MIL informs the driver that a fault that affects the vehicle's emission levels has occurred. The owner should take the vehicle in for service as soon as possible. The MIL will illuminate if a component or system that has an impact on emissions indicates a malfunction or fails to pass an emissions-related diagnostic test. The MIL will stay lit until the system or component passes the same test for three consecutive trips with no emissions-related faults. After making the repair, technicians may need to take the vehicle for three trips to ensure the MIL does not illuminate again.

As a bulb and system check, the MIL comes on with the ignition switch ON and the engine OFF. When the engine is started, the MIL turns off if there are no DTCs set.

When the MIL remains on while the engine is running, or when a malfunction is suspected due to a drivability or emissions problem, perform a system check. These checks expose faults a technician may not detect if other diagnostics are performed first.

If the vehicle is experiencing a malfunction that may cause damage to the catalytic converter, the MIL will flash once per second. If the MIL flashes, the driver needs to get the vehicle to the service department now. If the driver reduces speed or load and the MIL stops flashing, a code is set and the MIL stays on. This means the conditions that presented potential problems to the converter have passed with the changing operating conditions.

Possible causes for OBD-II AIR monitor failure are faulty secondary AIR solenoid and/or relay; damaged, loose, or disconnected wiring in the secondary air solenoid and/or relay circuit; defective aspirator valve; disconnected or damaged AIR hoses and/or tubes; a defective electric or mechanical air pump; air pump drive belt missing; and a faulty AIR check valve.

Task B.3 Identify failed exhaust gas recirculation (EGR) system; determine root cause of failure; determine needed repair.

Manufacturers calibrate the amount of EGR gas flow for each engine. If there is too much or too little, it can cause performance problems by changing the engine breathing characteristics. Also, with too little EGR flow, the engine can overheat, detonate, and emit excessive amounts of NO_x. When any of these problems exist and it seems likely that the EGR system is at fault, check the system.

A rough idle could possibly be caused by a stuck open EGR valve, a PVS that fails to open, dirt on the valve seat, or loose mounting bolts. (This also causes a vacuum leak and a hissing noise.) A no-start condition, surging, or stalling can be caused by an open EGR valve. Detonation or spark knock can be caused by any condition that prevents proper EGR gas flow, such as a valve stuck closed, leaking valve diaphragm, restrictions in flow passages, EGR disconnected, or a problem in the vacuum source. Excessive NO_x emissions can be caused by any condition that prevents the EGR from allowing the correct amount of exhaust gases into the cylinder or anything that allows combustion temperatures. Poor fuel economy is typically caused by the EGR system if it relates to detonation or other symptoms of restricted or zero EGR flow.

Before attempting to troubleshoot or repair a suspected EGR system on a vehicle, make sure the engine is mechanically sound, the injection system is operating properly, and the spark control system is working properly.

Most often an electronically controlled EGR valve functions in the same way as a vacuum-operated valve. Apart from the electronic control, the system can have all of the problems of any EGR system. Those that are totally electronic and do not use a vacuum signal can have the same problems as others, with the exception of vacuum leaks and other vacuum-related problems. Sticking valves, obstructions, and loss of vacuum produces the same symptoms as on non-electronic controlled systems. If an electronic control component is not functioning, the condition is usually recognized by the PCM.

The EGRV and EGRC solenoids, or the EVR, should normally cycle on and off frequently when EGR flow is being controlled (warm engine and cruise rpm). If they do not, it indicates a problem in the electronic control system or the solenoids. Generally, an electronic control failure results in low or zero EGR flow and might cause symptoms like overheating, detonation, and power loss.

On many engines, the EGR valve can be checked with a hand-operated vacuum pump. Before proceeding with this test, make sure the engine produces enough vacuum to properly operate the valve. This is done by connecting a vacuum gauge to the engine. Make sure it is a manifold vacuum source. Then start the engine and gradually increase speed to 2,000 rpm with the transmission in neutral. The reading should be above 16 in. Hg. If not, there could be a vacuum leak or exhaust restriction. Before continuing to test the EGR, check the MAP and/or correct the problem of low vacuum.

To check an EGR valve with a vacuum pump, remove the vacuum supply hose from the EGR valve port. Connect the vacuum pump to the port and supply 18 inches of vacuum. Observe the EGR diaphragm movement. When the vacuum is applied, the diaphragm should move. If the valve diaphragm did not move or did not hold the vacuum, replace the valve.

With the engine at normal operating temperature, check the vacuum supply hose to make sure there is no vacuum to the EGR valve at idle. Then plug the hose. On EFI engines, disconnect the throttle air bypass valve solenoid. Then observe the engine's idle speed. If necessary, adjust idle speed to the emission decal specification. Slowly apply 5 to 10 inches of vacuum to the EGR valve. The idle speed should drop more than 100 rpm (the engine may stall) and then return to normal again when the vacuum is removed. If the idle speed does not respond in this manner, remove the valve and check for carbon in the passages under the valve. Clean the passages as required or replace the EGR valve. Carbon may be cleaned from the lower end of the EGR valve with a wire brush, but do not immerse the valve in solvent, and do not sandblast the valve.

Although most testing of EGR valves involves the valve's ability to open and close at the correct time, we are not really testing what the valve was designed to do–control NO_x emissions. EGR systems should be tested to see if they are doing what they were designed to do.

Many technicians wrongly conclude by thinking an EGR valve is working properly if the engine stalls or idles very rough when the EGR valve is opened. Actually this test just shows that the valve was closed and it will open. A good EGR valve opens and closes, but it also allows the correct amount of exhaust gas to enter the cylinders. EGR valves are normally closed at idle and open at approximately 2,000 rpm. This is where the EGR system should be checked.

To check an EGR system, use a five-gas exhaust analyzer. Allow the engine to warm up, and then raise the engine speed to around 2,000 rpm. Watch the NO_x readings on the analyzer. The meter measures NO_x in parts per million. In most cases, NO_x should be below 1,000 ppm. It is normal to have some temporary increases over 1,000 ppm; however, the reading should be generally less than 1,000. If the NO_x is above 1,000, the EGR system is not doing its job. The exhaust passage in the valve is probably clogged with carbon.

If only a small amount of exhaust gas is entering the cylinder, NO_x will still be formed. A restricted exhaust passage of only 1⁄8 inch will still cause the engine to run rough or stall at idle, but it is not enough to control combustion chamber temperatures at higher engine speeds. Keep in mind: Never assume the EGR passages are okay just because the engine stalls at idle when the EGR is fully opened.

When the EGR valve checks out and visually everything looks fine but a problem with the EGR system is evident, test the EGR controls. Often a malfunctioning electronic control will trigger a DTC. Service manuals give the specific directions for testing these controls; always follow them.

OBD-II systems will perform certain tests on various subsystems of the engine management system. OBD-II is designed to turn on the MIL when the vehicle has any failure that could potentially cause the vehicle to exceed its designed emission standard by a factor of 1.5. The system does that by the use of a monitor. If one or more monitored systems are out limit, then the MIL turns on to indicate a problem. The EGR system is one of those systems monitored by OBD-II.

The EGR monitors use several different strategies to determine if the system is operating properly. Some monitor the temperature within the EGR passages. A high temperature indicates that exhaust gas is present. Other systems look at the MAP signal, energize the EGR valve, and look for corresponding change in vacuum levels.

The EGR system may contain a delta pressure feedback EGR (DPFE) sensor. An orifice is located under the EGR valve, and small exhaust pressure hoses are connected from each side of this orifice to the DPFE sensor. During the EGR monitor, the PCM first checks the DPFE signal. If this sensor signal is within the normal range, the monitor proceeds with the tests.

With the engine idling and the EGR valve closed, the PCM checks for pressure difference at the two pressure hoses connected to the DPFE sensor. When the EGR valve is closed and there is no EGR flow, the pressure should be the same at both pipes. If the pressure is different at these two hoses, the EGR valve is stuck open.

The PCM commands the EGR valve to open and then checks the pressure at the two exhaust hoses connected to the DPFE sensor. With the EGR valve open and EGR flow through the orifice, there should be higher pressure at the upstream hose than at the downstream hose.

The PCM checks the EGR flow by checking the DPFE signal value against an expected DPFE value for the engine operating conditions at steady throttle within a specific rpm range. If a fault is detected in any of the EGR monitor tests, a DTC is set in the PCM memory. If the fault occurs during two drive cycles, the MIL light is illuminated. The EGR monitor operates once per OBD-II trip.

The MIL is on the instrument panel. The MIL informs the driver that a fault that affects the vehicle's emission levels has occurred. The owner should take the vehicle in for service as soon as possible. The MIL will illuminate if a component or system that has an impact on emissions indicates a malfunction or fails to pass an emissions-related diagnostic test. The MIL will stay lit until the system or component passes the same test for three consecutive trips with no emissions-related faults. After making the repair, technicians may need to take the vehicle for three trips to ensure the MIL does not illuminate again.

If the vehicle is experiencing a malfunction that may cause damage to the catalytic converter, the MIL will flash once per second. If the MIL flashes, the driver needs to get the vehicle to the service department, now. If the driver reduces speed or load and the MIL stops flashing, a code is set and the MIL stays on. This means the conditions that presented potential problems to the converter have passed with the changing operating conditions.

For the catalyst efficiency, HO_2S, EGR, and comprehensive component monitors, the MIL is turned off if the same fault does not reappear for three consecutive drive cycles. When the fault is no longer present and the MIL is turned off, the DTC is erased after 40 engine warmup cycles. A technician may use a scan tester to erase DTCs immediately. A pending DTC is a code representing a fault that has occurred but that has not occurred enough times to illuminate the MIL. Some scan testers are capable of reading pending DTCs with the continuous DTCs.

Possible causes of OBD-II EGR monitor failure are faulty EGR valve; faulty EGR passages or tubes; loose or damaged EGR solenoid wiring and/or connectors; damaged DPFE or EGR VP sensor; disconnected or loose electrical connectors to the DPFE or EGR VP sensors; and disconnected, damaged, or misrouted EGR vacuum hoses.

Task B.4 Identify failed early fuel evaporation (EFE) system [heat riser]; determine root cause of failure; determine needed repair.

Hydrocarbon and carbon monoxide exhaust emissions are highest when the engine is cold. The introduction of warm combustion air improves the vaporization of the fuel in the carburetor, fuel injector body, or intake manifold. On older engines equipped with a carburetor or TBI unit, the fuel is delivered above the throttles, and the intake manifold is filled with a mixture of air and gasoline vapor. In these applications, some intake manifold heating is required to prevent fuel condensation, especially when the intake manifold is cool or cold. Therefore, these engines have intake manifold heat control devices such as heated air inlet systems, manifold heat control valves, and early fuel evaporation (EFE) heaters.

A heated air inlet control is used on gasoline engines with carburetion or central fuel injection. This system controls the temperature of the air on its way to the carburetor or fuel injection body. By warming the air, it reduces HC and CO emissions by improved fuel vaporization and faster warmup.

Another system uses an exhaust manifold heat control valve that routes exhaust gases to warm the intake manifold when the engine is cold. This heats the air-fuel mixture in the intake manifold and improves ventilation. The result is reduced HC and CO emissions. These control valves can be either vacuum or thermostatically operated.

Some V8 engines use a more complicated manifold heat control valve, which is called a power heat control valve. It was designed to work with a mini-catalyst and to preheat the air-fuel mixture for improved cold engine drivability. A vacuum actuator keeps the power heat control valve closed during warmup. All right-side exhaust gas travels up through the intake manifold crossover to the left side of the engine. Then all exhaust gas from the engine passes through a mini-converter just down from the left manifold. This converter warms up rapidly because it is small and close to the engine. Its rapid warmup reduces exhaust emissions. As the engine and main converter warm up, a coolant-controlled engine vacuum switch closes. This cuts vacuum to the actuator and allows the valve to open. Exhaust gas flows through both manifolds into the exhaust system and main converter.

Task B.5 Identify failed oxygen sensor(s) (O_2S/HO_2S) component(s) and circuitry; determine cause of failure; determine needed repair.

The exhaust gas oxygen sensor's signal is used by the computer to maintain a balanced air-fuel mixture. The O_2S is threaded into the exhaust manifold or into the exhaust pipe near the engine.

One type of oxygen sensor, made with a zirconium dioxide element, generates a voltage signal proportional to the amount of oxygen in the exhaust gas. It compares the oxygen content in the exhaust gas with the oxygen content of the outside air. As the amount of unburned oxygen in the exhaust gas increases, the voltage output of the sensor drops. Sensor output ranges from 0.1 volt (lean) to 0.9 volt (rich). A perfectly balanced air-fuel mixture of 14.7:1 produces an output of around 0.5 volt. When the sensor reading is lean, the computer enriches the air-fuel mixture to the engine. When the sensor reading is rich, the computer leans the air-fuel mixture.

Because the oxygen sensor must be hot to operate, most late-model engines use heated oxygen sensors (HO_2S). These sensors have an internal heating element that allows the sensor to reach operating temperature more quickly and to maintain its temperature during periods of idling or low engine load.

A second type of oxygen sensor, a titania-type, does not generate a voltage signal. Instead it acts like a variable resistor, altering a base voltage supplied by the control module. When the air-fuel mixture is rich, sensor resistance is low. When the mixture is lean, resistance increases. Variable-resistance oxygen sensors do not need an outside air reference. This eliminates the need for internal venting to the outside. They feature very fast warmup, and they operate at lower exhaust temperatures.

The accuracy of the oxygen sensor reading can be affected by air leaks in the intake or exhaust manifold. A misfiring spark plug that allows unburned oxygen to pass into the exhaust also causes the sensor to give a false lean reading.

Before testing an O_2S, refer to the correct wiring diagram to identify the terminals at the sensor. Most late-model engines use heated oxygen sensors (HO_2S). These sensors have an internal heater that helps to stabilize the output signals. Most heated oxygen sensors have four wires connected to them. Two are for the heater and the other two are for the sensor.

An O_2S can be checked with a voltmeter. Connect it between the O_2S wire and ground. The sensor's voltage should be cycling from low voltage to high voltage. The signal from most oxygen sensors varies between 0 and 1 volt. If the voltage is continually high, the air-fuel ratio may be rich or the sensor may be contaminated. When the O_2S voltage is continually low, the air-fuel ratio may be lean, the sensor may be defective, or the wire between the sensor and the computer may have a high-resistance problem. If the O_2S voltage signal remains in a mid-range position, the computer may be in open loop or the sensor may be defective.

The activity of an O_2S is best monitored with a labscope. The scope is connected to the sensor in the same way as a voltmeter. The switching of the sensor should be seen as the sensor signal goes to lean to rich to lean continuously.

The activity of the sensor can also be monitored on a scanner. By watching the scanner while the engine is running, the O_2 voltage should move to nearly 1 volt, then drop back to close to zero volts. Immediately after it drops, the voltage signal should move back up. This immediate cycling is an important function of an O_2S. If the response is slow, the sensor is lazy and should be replaced. With the engine at about 2,500 rpm, the O_2S should cycle from high to low 10 to 40 times in 10 seconds. When testing the O_2S, make sure the sensor is heated and the system is in closed loop.

Computer systems with OBD-II capabilities have monitoring systems and strategies for several important emission-related subsystems. The O_2S is looked at by several of these monitors.

OBD-II vehicles use a minimum of two oxygen sensors. One of these is used for feedback to the PCM for fuel control, and the other, located at the rear of the catalytic converter, gives an indication of the efficiency of the converter. The downstream O_2S is sometimes called the "catalyst monitor sensor" (CMS).

If the converter is operating properly, the signal from the pre-catalyst O_2S will have oscillations, while the post-catalyst O_2S will be relatively flat. Once the signal from the rear sensor approaches that of the front sensor, the MIL comes on and a DTC is set.

The downstream heated oxygen sensors have additional protection to prevent the collection of condensation on the ceramic. The internal heater is not turned on until the ECT sensor signal indicates a warmed-up engine. This action prevents cracking of the ceramic. Gold-plated pins and sockets are used in the heated oxygen sensors, and the downstream and upstream sensors have different wiring harness connectors.

A catalytic converter stores oxygen during lean engine operation and gives up this stored oxygen during rich operation to burn up excessive hydrocarbons. Catalytic converter efficiency is measured by monitoring the oxygen storage capacity of the converter during closed-loop operation.

When the catalytic converter is storing oxygen properly, the downstream HO_2S provide low-frequency voltage signals. If the catalytic converter is not storing oxygen properly, the voltage signal frequency increases on the downstream HO_2S until the frequency of the downstream HO_2S approaches the frequency of the upstream HO_2S. When the downstream HO_2S voltage signals reach a certain frequency, a DTC is set in the PCM memory. If the fault occurs on three drive cycles, the MIL light is illuminated.

There is also a heated oxygen sensor monitor, which monitors lean-to-rich and rich-to-lean time responses. This test can pick up a lazy O_2S that cannot switch fast enough to keep proper control of the air-fuel mixture in the system. These sensors are the heated type, and the amount of time before activity of the sensor signal is present is an indication of whether it is functional or not. Some systems use current flow to indicate if the heater is working or not.

All of the system's HO_2S sensors are monitored once per drive cycle; but the heated oxygen sensor monitor provides separate tests for the upstream and downstream sensors. The heated oxygen sensor monitor checks the voltage signal frequency of the upstream HO_2S. Excessive time between signal voltage frequencies indicates a faulty sensor. At certain times, the heated oxygen sensor monitor varies the fuel delivery and checks for HO_2S response. A slow response in the sensor voltage signal frequency indicates a faulty sensor. The sensor signal is also monitored for excessive voltage.

The heated oxygen sensor monitor also checks the frequency of the rear HO_2S signals and checks these sensor signals for excessively high voltage. If the monitor does not detect signal voltage frequency within a specific range, the rear HO_2S are considered faulty. The heated oxygen sensor monitor will command the PCM to vary the air-fuel ratio to check the rear HO_2S sensor response.

An OBD-II system continuously monitors the entire emissions system, switches on a MIL if something goes wrong, and stores a fault code in the PCM when it detects a problem. The codes are well defined and can lead a technician to the problem. A scan tool must be used to access and interpret emission-related DTCs regardless of the make and model of the vehicle.

When the same fault has been detected during two drive cycles, a DTC is stored in the PCM memory. If a misfire occurs, the misfire monitor will store a DTC immediately in the PCM memory, depending on the type of misfire. If a misfire that threatens engine or catalyst damage occurs, the misfire monitor flashes the MIL on the first occurrence of the misfire. A fault detected by the catalyst monitor must occur on three drive cycles before the MIL is illuminated.

For the misfire and fuel system monitors, if the fault does not occur on three consecutive drive cycles under similar conditions, the MIL is turned off. For the catalyst efficiency, HO_2S, EGR, and comprehensive component monitors, the MIL is turned off if the same fault does not reappear for three consecutive drive cycles. When the fault is no longer present and the MIL is turned off, the DTC is erased after 40 engine warmup cycles. A technician may use a scan tester to erase DTCs immediately.

Possible causes for OBD-II oxygen sensor monitor failure are malfunctioning upstream and/or downstream oxygen sensors, malfunctioning heater for the upstream or downstream oxygen sensor, a faulty PCM, and defective wiring to and/or from the sensors.

If the HO_2S wiring, connector, or terminal is damaged, replace the entire oxygen sensor assembly. Do not attempt to repair the assembly. In order for this sensor to work properly, it must have a clean air reference. The sensor receives this reference from the air that is present around the sensor's signal and heater wires. Any attempt to repair the wires, connectors, or terminals could result in the obstruction of the air reference and degraded oxygen sensor performance.

Additional guidelines for servicing a heated oxygen sensor follow: Do not apply contact cleaner or other materials to the sensor or wiring harness connectors. These materials may get into the sensor, causing poor performance. The sensor pigtail and harness wires must not be damaged in such a way that the wires inside are exposed. This could provide a path for foreign materials to enter the sensor and cause performance problems. Neither the sensor nor vehicle lead wires should be bent sharply or kinked. Hard bends, kinks, etc., could block the reference air path through the lead wire. Do not remove or defeat the oxygen sensor ground wire. Vehicles that utilize the ground wired sensor may rely on this ground as the only ground contact to the sensor. Removal of the ground wire will cause poor engine performance. To prevent damage due to water intrusion, be sure the peripheral seal remains intact on the vehicle harness.

Task B.6 Inspect emission systems for evidence of tampering (missing/modified and/or improperly installed components); determine needed repair.

The Environmental Protection Agency (EPA) establishes emissions standards that limit the amount of these pollutants a vehicle can emit. To meet these standards, many changes have been made to the engine itself. Plus there have been systems developed and added to the engines to reduce the pollutants they emit. A list of the most common pollution control devices related to the exhaust system follows.

- *Exhaust gas recirculation (EGR) system.* This system introduces exhaust gases into the intake air to reduce the temperatures reached during combustion. This reduces the chances of forming NO_x during combustion.
- *Catalytic converter.* Located in the exhaust system, it allows for the burning or converting of HC, CO, and NO_x into harmless substances, such as water.
- *Air-injection system.* This system reduces HC emissions by introducing fresh air into the exhaust stream to cause minor combustion of the HC in the engine's exhaust.

It is against the law for anyone to tamper, remove, or intentionally damage any part of the emission control system. When inspecting an exhaust system, carefully look at the system and attempt to identify any signs of tampering. Sometimes the law violations are obvious, such as the presence of a straight pipe where the catalytic converter should be. Some tampering is not as quickly seen, such as the enlargement of the fuel filler neck so that it can accept leaded fuels. Use of leaded fuels is deadly to the catalytic converter. Also check the piping to the AIR and EGR systems. Any evidence of tampering (missing/modified and/or improperly installed components), must be noted prior to doing any repair work and must be corrected during the repair procedure. Failure to follow these guidelines would make you liable for any and all fines levied against the owner of the vehicle.

If a catalytic converter is found to be bad, under no circumstances should the converter be removed and replaced with a straight piece of pipe (a test pipe). This practice is illegal for the professional auto technician. Because of constant change in EPA catalytic converter removal and installation requirements, check with the manufacturer or EPA for the latest data regarding replacement.

Replacement converters must have documents or labels that show they meet EPA requirements and are warranted to meet federal durability and performance standards. All manufacturers of new and rebuilt converters who meet the EPA requirements must state that fact in writing, usually in the warranty information in their catalogs. If you bypass a damaged or ineffective catalytic converter instead of replacing it or repairing it, you are breaking the law! The penalty could be much higher than the price of a new or rebuilt unit.

C. Exhaust System Fabrication (6 Questions)

Task C.1 Pipe Bending (3 Questions)

Task C.1.1 Determine bending method (program card, pattern/copy, or custom).

It is important that an exhaust system fit properly under the vehicle to ensure rattle-free operation. When one considers all of the various models of vehicles on the road, each with their own unique exhaust system, it is easy to see why fabrication of a system may be a necessary duty of an undercar specialist.

Exhaust system parts can be obtained from a supplier or by keeping a supply of parts in the shop. However, stocking enough exhaust pipes, mufflers, and tailpipes takes much space and involves a large amount of money. To help overcome these problems, tube-bending equipment is available. With this equipment, only straight tubing in various diameters is stocked. The straight pieces are then bent to the desired shape.

Many tube-bending tools are fully automatic and are designed to produce bends through 3 inches outside diameter tubing. A heavy-duty expander is used on the ends of the cut to form slip connections.

When bending pipe for an exhaust system, the pipe can be bent to match the old or original pattern. To do this, the straight pipe is laid out against the original and bends are made one at a time until the pipe matches the original. This is a time-consuming skill as each bend is compared and tweaked until it matches the original. Some pipe-bending machines are somewhat intelligent. These use program cards to bend the pipe at preset angles for the vehicle. Again the pipe is bent one bend at a time, until it matches the original configuration. It is important to match the location and twist of the bends.

Having the correct bends is not enough; the bends must be at the right place and at the right angle. Sometimes an undercar specialist will need to make a totally custom exhaust system. This may be due to changes made to the vehicle or by customer request. In these cases, the exhaust system needs to be carefully planned out and each step checked.

In all cases, the exhaust system needs to be checked for clearances before tightening it in place.

Task C.1.2 Determine center of bends, rotation of pipe, depth of bends, and pipe diameter(s); perform bending operation.

When bending pipe for an exhaust system, the pipe can be bent to match the old or original pattern. To do this, the straight pipe is laid out against the original and bends are made one at a time until the pipe matches the original. This is a time-consuming skill as each bend is compared and tweaked until it matches the original. Some pipe-bending machines are somewhat intelligent. These use program cards to bend the pipe at preset angles for the vehicle. Again the pipe is bent one bend at a time, until it matches the original configuration. It is important to match the location and twist of the bends. Having the correct bends is not enough; the bends must be at the right place and at the right angle.

When making a bend in the pipe, make sure the depth of the bend is exactly what is required. Then using the original pipe as a guide, rotate the pipe to position the pipe for its next bend. Work pipe from one end to the other. Do not haphazardly bend the pipe.

After all of the bends have been made, it may be necessary to tweak the pipe to ensure proper placement. Sometimes an Undercar Specialist will need to make a totally custom exhaust system.

This may be due to changes made to the vehicle or by customer request. In these cases, the exhaust system needs to be carefully planned out and each step checked.

In all cases, the exhaust system needs to be checked for clearances before tightening it in place.

When installing a pipe into a pipe-bending tool, make sure to follow the tool manufacturer's instructions. It is important that the center part of the bend be in the center of the pipe holding fixture. Also make sure the ends of the pipe are properly secured. Set the machine to the desired bend or vehicle and begin the bending process.

Task C.1.3 Perform end-forming and hardware installation operations.

Once the pipe is shaped, it needs to be cut to the correct length. Before doing this, determine the needed amount of overlap for joints. Make sure the cuts are straight and even. It may be necessary to ream the inside of the pipe to remove any depression that may have resulted from the cut.

The ends of an exhaust pipe must be prepared to join with other components or to serve as a tailpipe. Tools are available to properly form a flange. These flaring tools are typically built to accommodate a variety of pipe diameters. Make sure you choose the correct tools for the pipe on which you are working. Also, it is wise to place any flange sealing hardware over the pipe before flaring the end. Check the flare by inserting a new gasket into the flare to make sure the flare is round and will provide a good sealing area for the gasket.

If the end of a pipe must form a slip joint with another component, use the correct pipe expander to accomplish the female end. Make sure the expansion is just enough to allow the other component to fit in. If the pipe is excessively expanded, it may be difficult to form a good seal. Also, excessive expansion may weaken the pipe.

Task C.1.4 Determine the cause of pipe material failures that occur during bending operations.

Task C.2 Welding and Cutting (3 Questions)

Task C.2.1 Select appropriate welding method (gas or MIG); perform welding operation; verify integrity of weld.

Exhaust system pipes can be welded together. Welding, of course, is a process in which heat is used to melt metal and allow the molten portions to flow together. When cooled, a welded joint is as strong as, or sometimes stronger than, a single piece of metal. Do not attempt to use welding equipment without proper training and safety equipment. The light caused by welding can cause permanent eye damage or even blindness.

Furthermore, the heat from welding can ignite fuel vapors or damage parts. Oxyacetylene welding provides a high-temperature flame for welding by combining oxygen and acetylene gases. The flame of this combination provides enough heat to melt most metals. Oxyacetylene welding is a manual process; the welder controls the torch movement and the welding rod. The torch is connected via hoses to separate cylinders of gas. The flow rate from the cylinders to the torch is controlled by a manifold gauge set at the tanks and by needle valves at the torch. One torch valve controls the rate of oxygen flow and the other controls the rate of acetylene flow. The two gases mix and burn at the orifice of the torch.

Oxyacetylene welding joins metals (whether the metals are two components or a component and a welding rod) by melting and fusing. A very intense, concentrated flame is applied to the metal until a spot under the flame becomes molten and forms a pool of liquid metal. When two metals melt and the molten pools run together and fuse, the edges of the two become one.

Gas metal arc welding uses an electric arc between a continuously fed metal electrode and the base metal to produce heat. This arc is shielded by a gas and is therefore commonly referred to as MIG or metal inert gas welding. A gas cylinder, regulator, flow meter, and hose provide a flow of the shielding gas to the arc. Carbon dioxide, argon, or helium can be used as this shielding gas. A cable carries the electrode wire, current, and shielding gas to the torch and arc. The torch usually has a trigger-type switch for starting and stopping the electrode feed and gas flow. A constant voltage DC welder is used and the desired voltage is set on the welding machine. Current is changed by adjusting the feed speed of the wire.

The metal to be welded determines what welding method should be used. MIG can be used on all metals; whereas oxyacetylene welding is not typically used with stainless steel and must not be used on high-strength steel or aluminum. The work area is also something that needs to be considered when deciding what welding method to use.

When using a MIG welder or any other type of electric welder on a vehicle, it is important that you disconnect the battery and that the vehicle's computers are totally isolated. Failure to do so may cause damage to the computers and other electronic components.

Perform welding operation; verify integrity of weld.

Oxyacetylene welding begins with the proper setup of the equipment and then lighting the torch. It is important that the correct torch tip and rod be selected before proceeding. Visually inspect the condition of the equipment. If anything looks potentially hazardous, do not proceed. Turn the regulator adjusting screws all the way out before opening the cylinder valves. This prevents damage to the regulator.

Stand to one side of the regulator when opening the cylinder valves. The high pressure can cause a weak or damaged regulator to burst and cause injury. Slowly open the acetylene cylinder valve one-quarter to one-half turn. Open the oxygen valve very slowly. When the regulator high-pressure gauge reaches its highest reading, turn the cylinder valve all the way open. Then open the acetylene torch valve one turn. Turn the acetylene regulator adjusting screw in slowly until the low-pressure acetylene gauge indicates a pressure that is correct for the torch tip. Then close the acetylene torch valve.

Open the torch oxygen valve one turn. Turn the oxygen regulator adjusting screw in until the low-pressure oxygen gauge indicates the correct pressure for the torch tip. Then close the oxygen torch valve.

Purge the lines independently by cracking open the torch valves. After purging, open the acetylene torch valve slightly. Use an igniter to light the acetylene gas at the torch's tip. Continue to open the acetylene torch valve until the flame slightly jumps away from the end of the tip.

After the acetylene is regulated, slowly open the oxygen valve on the torch. As the oxygen is fed into the flame, the acetylene flame will become purple and a small inner cone will begin to form. This inner cone is white and will become whiter as more oxygen is added. As oxygen flow increases, the inner cone becomes a round, smooth cone. Stop the adjustment at this point. When correctly adjusted, the flame will be neutral and emit a soft purring noise.

To weld, move the torch in the direction in which the tip is pointing. The torch should be held at an angle of 15 to 75 degrees. The angle depends on the tip size, the thickness of the metal, and the welding conditions. The flame will spread over the work ahead of the weld to preheat the metal before it comes under the high-temperature flame. After a weld pool is evident, the welding rod can be held in the area and welding begins.

To shut off the torch, close the acetylene torch valve and then the oxygen torch valve. Tightly close the cylinder valves and reopen the torch valves to bleed the hoses. Bleed until the gauges read zero, then lightly close the torch valves.

Using MIG welding, the welder must select the size of the electrode, then set the unit to the desired voltage. The welder also must adjust the flow of the shielding gas and the rate of electrode feed.

Before welding, check the equipment to make sure it is safe to use. Prior to filling in a weld, tack the parts into their desired position. The arc length is determined by the voltage; therefore, the welder must watch and control the distance from the nozzle to the work. By controlling the nozzle-to-work distance, the welder will control the electrode extension distance. The welding speed and torch angle determines the bead width and appearance.

To start welding, tip the top of the torch 5 to 15 degrees in the direction of travel and place the helmet down over your eyes. To start the arc, the wire feeder and the gas, squeeze the trigger of the gun. The wire will arc as soon as it feeds out far enough to touch the metal. As the weld pool reaches a proper width, the gun is moved forward along the weld. A steady and consistent motion will ensure a constant and uniform bead. More than one pass may be required to fill a weld groove. Each pass should be cleaned before the next pass is laid.

Some common faults that must be avoided in welding are: (1) too much penetration, (2) lack of penetration, (3) slag inclusions, (4) porosity, (5) lack of fusion, and (6) undercutting.

Task C.2.2 Set up and adjust welding equipment to repair application; observe applicable personnel, vehicle, and equipment safety procedures.

Loose clothing, unbuttoned shirtsleeves, dangling ties and jewelry, and shirts hanging out are very dangerous in a shop; instead, wear approved shop coveralls or jumpsuits. Pants should always be long enough to cover the top of the shoes. This will prevent sparks from going down into the shoes, especially when you are using welding equipment. For added welding safety, welder's pants, leggings, or spats are often worn. A welding filter lens, sometimes called a filter plate, is a shaded glass welding helmet insert used to protect your eyes from ultraviolet burns. The lenses are graded from 4 to 12. The higher the number, the darker the filter. The American Welding Society (AWS) recommends grade #9 or #10 for MIG welding steel. Oxyacetylene welding should be done with a #4, #5, or #6 tinted filter shade.

In MIG welding, the polarity of the power source is important in determining the penetration to the work piece. Direct current (DC) power sources used for MIG welding typically use DC reverse polarity. DC reverse polarity means the wire (electrode) is positive, and the work piece is negative. Weld penetration is greater using this connection. Weld penetration is also greatest using CO_2 gas. However, CO_2 gives a harsher, more unstable arc, which leads to increased spatter. So when welding on thin materials, it is preferable to use argon/carbon dioxide. Voltage adjustment and wire feed speed must be set according to the diameter of the wire being used and metal thickness. It should be noted that when setting these parameters, follow the manufacturer's recommendations to reach approximate settings.

When the arc voltage is low, the arc length decreases, penetration is deep, and the bead is narrow and dome-shaped. When the clamp is attached to clean metal on the vehicle near the weld site, it completes the welding circuit from the machine to the work and back to the machine. This clamp is not referred to as a ground cable or ground clamp. The ground connection is for safety purposes and is usually made from the machine's case to the building ground through the third wire in the electric input cable.

Attach the clamp to a clean metal surface to prevent electrical resistance that can affect weld current. Welding current affects the base metal penetration depth, the speed at which the wire is melted, arc stability, and the amount of weld spatter. As the electrical current is increased, the penetration depth, excess metal height, and bead width also increase. Too much penetration causes burn-through beneath lower base metal.

Handle the cylinder of shielding gas with care. It might be pressurized to more than 2,000 psi (13,790 kPa). Chain or strap the cylinder to a support sturdy enough to hold it securely to the MIG machine. Install the regulator, making sure to observe the recommended safety precautions. Regulators reduce the pressure coming from the tanks to the desired level and maintain a constant flow rate of oxygen pressure (15 to 100 psi) and acetylene (3 to 12 psi).

Overhead welding has the work pieces turned upside down. Overhead welding is difficult. In this position, the danger of having too large a puddle is obvious; some of the molten metal can fall down into the nozzle, where it can create problems. Thus, always do overhead welding at a lower voltage while keeping the arc as short as possible and the weld puddle as small as possible. Press the nozzle against the work to ensure that wire is not moved away from the puddle. It is best to pull the gun along the joint with a steady drag.

The results from welding with oxyacetylene are determined by the mixture of the gases. This mixture is indicated by the condition of the flame. A carburizing flame is blue with an orange and red end. This results from excessive acetylene. A neutral flame has a quiet blue-white inner cone. A neutral flame is the result of the correct amount of acetylene and oxygen. An oxidizing flame has a short and hissing inner cone. This flame results from an excess of oxygen in the mixture and tends to burn the metal being welded.

Certain conditions are necessary for a good weld. The temperature of the flame must be great enough to melt the metals. There must also be enough heat to overcome any heat losses. The flame must not add dirt, foreign material, or carbon to the metal being worked. The flame must not burn or oxidize the metal. The products of combustion should not be toxic.

The amount of heat is determined by the amount of gas being burned. To obtain more heat, a torch tip with a larger orifice is used. To provide the right amount of gas to the larger orifice, higher pressures are set. Likewise, if low heat is needed, a torch tip with a smaller orifice is used with lower pressures. Regardless of the torch tip or the pressures used, if the flame is neutral it will have the same temperature regardless. More heat is present with a larger tip because the flame covers more area at one time.

If the torch tip orifice is too small, not enough heat will be available to bring the metal to its melting and flowing temperature. If the torch tip is too large, poor welds will result because the weld will need to be made too quickly, the welding rod will melt too quickly, and the weld pool will be hard to control.

When welding with oxyacetylene, make sure the hoses are not allowed to come in contact with any flame or hot metal. Also make sure the hoses are not sharply kinked. Kinking may cause the hose to crack, and the pressure in the line may cause the hose to burst. A kink will also decrease gas flow. When welding, the hose should be protected from falling objects and should not be stepped on.

A welding rod is filler metal that does not conduct electricity. Welding rods are commonly made of mild steel, cast iron, stainless steel, and brazing alloys. Welding rods are available in many different diameters. The diameter and type of rod used is determined by the strength needed at the weld and metal being welded.

Task C.2.3 Select appropriate cutting method (gas or mechanical); perform cutting operation.

When trying to replace a part in the exhaust system, you may run into parts that are rusted together. This is especially a problem when a pipe slips into another pipe or the muffler. If you are trying to reuse one of the parts, you should carefully use a cold chisel or slitting tool on the outer pipe of the rusted union. You must be careful when doing this, because you can easily damage the inner pipe. It must be perfectly round to form a seal with a new pipe.

Manual pipe cutters can be used to totally cut through an old pipe as can heat. The type of material being cut and the work environment are the primary considerations when deciding how to cut a pipe.

Oxyacetylene equipment should not be used on high-strength steel components for welding or cutting. Vehicle manufacturers' engineers stress this point. There is just too much heat buildup that can reduce structural strength. Plasma arc cutting is replacing oxyacetylene as the best way to cut metals. Plasma arc cutting cuts damaged metal effectively and quickly, but does not destroy the properties of the base metal.

When using a plasma arc cutter, be aware of the fact that the sparks from the arc can damage painted surfaces and can also pit glass. Use a welding blanket to protect these surfaces. Also, when welding or using a plasma arc cutter, make sure there is nothing behind the panel that can be damaged. Check for wiring, fuel lines, sound deadening materials, and other objects that can cause a fire. Disconnect the vehicle's battery; the open circuit voltage of plasma cutting equipment can be very high (in the range of 250 to 300 volts), so insulated torches and internally connected terminals are also essential.

With oxyfuel gas cutting, the flame raises the temperature of the metal to its melting point. A high-pressure jet of oxygen from the cutting torch is directed at the metal that causes the metal to burn and blow away very rapidly. The equipment needed to use acetylene for cutting is the same as for welding except a cutting torch is used in place of the welding torch.

Perform cutting operation. The cutting torch must be carefully used in order to provide clean and accurate cuts. The torch tip must be in good condition and set to align with the cutting line. Also, the preheat flame must be correctly adjusted and the cutting oxygen pressure must be correct.

To cut, bring the tip of the inner cone of the preheat flame to the edge of the metal to be cut. The torch should be held so that the inner cone of the preheat flame is about 1⁄16 to 1⁄8 inch away from the surface of the metal.

As soon as a spot on the cutting line has been heated to a bright cherry red color, squeeze the cutting oxygen lever. The jet of oxygen coming from the tip will cause the heated metal to burn away.

Task C.2.4 Set up and adjust cutting equipment to repair application; observe applicable personnel, vehicle, and equipment safety procedures.

With oxyfuel gas cutting, the flame raises the temperature of the metal to its melting point. A high-pressure jet of oxygen from the cutting torch is directed at the metal that causes the metal to burn and blow away very rapidly. The equipment needed to use acetylene for cutting is the same as for welding except a cutting torch is used in place of the welding torch.

Just as with welding, the proper size cutting tip is very important. The preheat flame must provide just the right amount of heat, and the oxygen jet orifice must deliver the correct amount of oxygen at the correct pressure and speed to provide a good cut. The cutting tip has one or more preheating orifices to provide the flames used to heat the metal. Cutting oxygen exits from a central orifice when the welder depresses the cutting oxygen lever. The oxyfuel gas preheating flame is adjusted in the same way as for welding. The cutting oxygen is adjusted to provide enough blow to cut, but not so much that the preheating flame is blown out.

With using oxyfuel for cutting, be sure your work area is well ventilated and clear of combustible materials. Remove any possible hazards. Check that the regulator adjustment screws are backed out until there is no tension against them. If they are not, back them out. Close both valves at the torch connection. Install the proper tip for work.

Do not overtighten the tip. Make sure all connections are tight. Open the oxygen cylinder valve slowly until the cylinder pressure gauge gives an indication. Then open the valve completely. Check for any leaks in the line or at the tip.

Open the acetylene cylinder valve slowly until the cylinder pressure gauge gives an indication. Then open the valve an additional half-turn. Check for leaks. Open the oxygen valve at the tip a half-turn. Adjust the regulator until a working pressure of 8-25 psi is indicated. The amount of working pressure will depend upon the type of work being performed. Close the valve.

Open the acetylene valve at the tip a half-turn. Adjust the working pressure to 3 to 8 psi. The amount of working pressure will depend on the work being performed. Close the valve. Make sure all safety equipment and protective clothing is in place. Open the acetylene valve a half-turn and ignite the gas. Continue to open the valve until the black smoke disappears. A reddish-yellow flame should appear.

Slowly open the oxygen valve at tip until a blue flame with a yellow-white cone appears. Continue to open the valve slowly until the center cone becomes sharp and well defined. This is the neutral flame.

Check the setting for the cutting oxygen. If the flame intensifies and roars without blowing out the flame, the setting is correct. If there is too much pressure, the preheat flame will blow out. If there is an insufficient amount of oxygen, the flame will not roar. Adjust the oxygen, according and recheck the preheat flame. Certain things should be considered whenever using an oxyacetylene torch:

- Cylinders must be secured to a wall or stationary structure when not in use or in a cart.
- Cylinders on a cart must be chained to the cart.
- When moving the cart, point the regulators away from your body.
- Wear proper filter-tinted eye protection.
- Cutting or welding fumes may be toxic. Work in a well-ventilated area, and use a respirator as needed.
- Do not weld or cut near combustible materials.

D. Exhaust System Installation (8 Questions)

Task D.1 Identify exhaust system configuration and options according to manufacturer's specifications (routing, single/dual, etc.)

Exhaust systems are designed for the vehicle they are in and the engine that is in the vehicle. In-line engines have a simple design that runs down one side of the engine. Called a single exhaust system, there is only one pipe to the rear of the car. Single exhausts can be found on all engine sizes. A single exhaust system has one path for exhaust flow through the system. Typically, it has only one header pipe, a main catalytic converter, a muffler, and a tailpipe. This is the most common type of exhaust system.

V-type engines with dual exhaust have two pipes, two mufflers, and two catalytic converters. A dual exhaust allows better breathing when the engine is under load. A dual-exhaust system has two separate exhaust paths to reduce backpressure. A crossover pipe joins the two manifolds into one pipe as they leave the engine on a V-type engine's single exhaust system. A crossover pipe normally connects the right and left header pipes to equalize backpressure in a dual system. This also increases engine power slightly.

Though most cars and light trucks have exhaust systems with only a single catalytic converter, there are some with dual converters or piggyback converters. Service and repair procedures are pretty much the same on the dual-converter systems as those with only a single converter.

Task D.2 Select components according to accepted standards regarding material, type, design, and size.

All the parts of the system are designed to conform to the available space of the vehicle's undercarriage and yet be a safe distance above the road. The muffler is a cylindrical or oval-shaped component, generally about 2 feet long, mounted in the exhaust system about midway or toward the rear of the car. Inside the muffler is a series of baffles, chambers, tubes, and holes to break up, cancel out, or silence the pressure pulsations that occur each time an exhaust valve opens.

The design of the muffler varies from one manufacturer to another. One design is the straight-through type. In this design, a straight path for the gases extends from the front to the rear of the unit. The path for the exhaust is through a centrally located pipe with holes. Sheet metal about three times the diameter of the pipe surrounds the pipe. Most often the muffler is filled with steel wool or some other heat-resistant, sound-deadening material.

Another type of muffler reverses the flow of the exhaust gases and has the advantage of saving space. The double-shell and two-shell designs are other forms of modern mufflers. There have been several important changes in recent years in the design of mufflers. Most of these changes have been centered at reducing weight and emissions, improving fuel economy, and simplifying assembly. More and more mufflers are being made of aluminum and stainless steel. Using these materials reduces the weight of the units as well as extending their lives. Retarded engine ignition timing that is used on many small cars tends to make the exhaust pulses sharper. Many cars use a double-wall exhaust pipe to better contain the sound and reduce pipe ring. Because the only space left under the car for the muffler is at the very rear of the vehicle, mufflers run cooler than before and they are more easily damaged by condensation in the exhaust system. This moisture, combined with nitrogen and sulfur oxides in the exhaust gas, forms acids that rot the muffler from the inside out. Many mufflers are being produced with drain holes drilled into them.

The design of a muffler is precise. The size and shape of the different chambers will affect the noise level and backpressure. Helmholtz tuning chambers within the muffler are precisely tuned. Chamber volume, tuning tube size, and temperature of the gases in the chamber are taken into account. The high-frequency tuning chamber reduces the sound level of the high frequencies present in the exhaust system. (Helmholtz chambers primarily affect low-frequency sounds.) The high frequencies can be generated by exhaust flow past a sharp edge in the exhaust system, venturi noise in the carburetor, and friction between the forceful exhaust flow and the pipes. High frequencies show up as a whistling noise. Therefore, each hole in the inner tube of the high-frequency tuning chamber acts as a small tuning tube. The reversing unit crossover is most effective in removing or reducing the mid-range frequencies missed by the high- and low-frequency chambers. The amount of crossover is determined by the size and number of holes in the adjacent tubes.

If the exhaust pipe is the right length and diameter, the frequency of the explosions can cause a resonance in that pipe. This is the same as blowing across the neck of a bottle. The Helmholtz tuning chambers can be designed to absorb resonance and reduce the noise level of the exhaust system.

Mufflers come in many inlet/outlet sizes and configurations. Some feature a center inlet and offset outlet, while some units have both the inlet and outlet offset. Be sure to check your application to determine the proper orientation. Inlet and outlet sizes also vary and you will need to make sure they match with the exhaust pipes. A common misconception is that very large diameter exhaust pipes are better. Typically, increasing pipe diameter by 1⁄4–1⁄2 inch over stock is sufficient to increase flow.

On some older vehicles, there is an additional muffler, known as a resonator or silencer. This unit is designed to further reduce or change the sound level of the exhaust. It is located toward the end of the system and generally looks like a smaller, rounder version of a muffler.

Many vehicles are equipped with a mini-catalytic converter that is either built into the exhaust manifold or is located next to it. These converters are used to clean the exhaust during engine warmup and are commonly called warmup converters. Many catalytic converters have an air hose connected from the AIR system to the oxidizing catalyst.

This air helps the converter work by making extra oxygen available. The air from the AIR system is not always forced into the converter; rather it is controlled by the vehicle's PCM. Fresh air added to the exhaust at the wrong time could overheat the converter and produce NO_x, something the converter is trying to destroy.

Except for quite old ones, all vehicles are fitted with a catalytic converter. Although converters are very effective in reducing emissions, they also are effective in reducing exhaust sound. This consequence is one of the leading factors in the design of new mufflers; they have less sound to reduce.

OBD-II regulations call for a way to inform the driver that the vehicle's converter has a problem and may be ineffective. The PCM monitors the activity of the converter by comparing the signals of an HO_2S located at the front of the converter with the signals from an HO_2S located at the rear. If the sensors' outputs are the same, the converter is not working properly and the malfunction indicator lamp (MIL) on the dash will light.

Make sure exhaust pipes are of the correct diameter to meet OE specifications. Pipes of a smaller diameter restrict flow and can create enough backpressure to cause drivability complaints and poor performance.

Heat shields are used to protect other parts from the heat of the exhaust system and the catalytic converter. They are usually made of pressed or perforated sheet metal. Heat shields trap the heat in the exhaust system, which has a direct effect on maintaining exhaust gas velocity. Clamps, brackets, and hangers are used to properly join and support the various parts of the exhaust system. These parts also help to isolate exhaust noise by preventing its transfer through the frame or body to the passenger compartment. Clamps help to secure the exhaust system parts to one another. The pipes are formed in such a way that one slips inside the other. This design makes a close fit. A U-type clamp usually holds this connection tight. Another important job of clamps and brackets is to hold pipes to the bottom of the vehicle. Some exhaust systems are a single unit in which the pieces are welded together by the factory. By welding instead of clamping the assembly together, carmakers save the weight of overlapping joints as well as that of clamps. Clamps and brackets must be designed to allow the exhaust system to vibrate without transferring the vibrations through the car.

There are many different types of flexible hangers available. Each is designed for a particular application. Some exhaust systems are supported by doughnut-shaped rubber rings between hooks on the exhaust component and on the frame or car body. Others are supported at the exhaust pipe and tailpipe connections by a combination of metal and reinforced fabric hanger. Both the doughnuts and the reinforced fabric allow the exhaust system to vibrate without breakage that could be caused by direct physical connection to the vehicle's frame.

Exhaust backpressure is the pressure developed in the exhaust system when the engine is running. High backpressure reduces engine power. A well-designed exhaust system should have low backpressure. Even a well-designed muffler will produce some backpressure in the system. Excessive backpressure caused by defects in a muffler or other exhaust system part can slow or stop the engine. However, a small amount of backpressure can be used intentionally to allow a slower passage of exhaust gases through the catalytic converter. This slower passage results in more complete conversion to less harmful gases. Also, no backpressure may allow intake gases to enter the exhaust. The size of the exhaust pipes, catalytic converter, and muffler contributes to exhaust backpressure. Larger pipes and a free-flowing muffler, for example, would reduce backpressure.

Mufflers, tailpipes, and exhaust pipes wear out due to corrosion. External rusting is due to rain, snow, and humidity. In some northern states, this external rusting is speeded up by the use of salt on icy roads. The greatest amount of corrosion occurs inside the exhaust system, mostly in the muffler. This is because a gallon of water is formed for every gallon of fuel burned. Acids are also formed in the combustion process. The acids and water combine to quickly rust the inside of the exhaust system.

Until the exhaust system has reached operating temperature, much of the moisture will condense on the cool surfaces and collect in the muffler. As the muffler becomes hot, the moisture evaporates. On short drives, the muffler will not get hot enough and corrosion will occur. To reduce corrosion, most manufacturers are using rust-resisting coatings and/or special alloys in the design of the mufflers and pipes. Stainless steel is used or a ceramic coating is applied to the inside of the muffler and pipes.

Task D.3 Install appropriate exhaust system components (mufflers, resonators, catalytic converters, pipes, and manifolds).

All the parts of the system are designed to conform to the available space of the vehicle's undercarriage and yet be a safe distance above the road. An exhaust system quiets engine operation and carries exhaust fumes to the rear of the vehicle. Typical parts include the following:

- Exhaust manifold–connects the cylinder head exhaust ports to header pipe
- Header pipe–steel tubing that carries exhaust gases from the exhaust manifold to the catalytic converter or muffler
- Catalytic converter–device that removes pollutants from engine exhaust
- Intermediate pipe–tubing sometimes used between the header pipe and muffler or catalytic converter and muffler
- Muffler–It is a metal chamber for damping pressure pulsations to reduce exhaust noise. Mufflers come in many inlet/outlet sizes and configurations as well as many sizes and shapes: oval, round, and even rectangle.
- Resonator–It is an additional muffler designed to further reduce or change the sound level of the exhaust. It is located toward the end of the system and generally looks like a smaller, rounder version of a muffler.
- Tailpipe–tubing that carries exhaust from the muffler to the rear of the car body
- Hangers–devices for securing the exhaust system to the underside of the car body
- Heat shields–metal plates that prevent exhaust heat from transferring into another object
- Exhaust system clamps–U-bolts for connecting parts of the exhaust system together

Make sure exhaust pipes are of the correct diameter to meet OE specifications. Pipes of a smaller diameter restrict flow and can create enough backpressure to cause drivability complaints and poor performance. Many header pipes are connected to the exhaust manifold by a spring-loaded coupling. The coupling compresses and holds the "doughnut" seal between the exhaust manifold and header pipe. This design allows the engine to move on its motor mounts without moving the exhaust system and damaging the seal.

Before beginning work on the system, be sure it is cool to the touch. Some technicians disconnect the battery ground to avoid short-circuiting the electrical system. Soak all rusted nuts, bolts, etc., in a good quality penetrating oil. Finally, check the system for critical clearance points so they can be maintained when new components are installed.

To replace a damaged exhaust pipe, begin by supporting the converter to keep it from falling. Carefully remove the oxygen sensor if there is one. Remove any hangers or clamps holding the exhaust pipe to the frame. Unbolt the flange holding the exhaust pipe to the exhaust manifold. When removing the exhaust pipe, check to see if there is a gasket. If so, discard it and replace it with a new one. Once the joint has been taken apart, the gasket loses its effectiveness. Disconnect the pipe from the converter and pull the front exhaust pipe loose and remove it.

An easy way to break off rusted nuts is to tighten them instead of loosening them. Sometimes a badly rusted clamp or hanger strap will snap off with ease. Sometimes the old exhaust system will not drop free of the body because a large part is in the way, such as the rear end or the transmission support. Use a large cold chisel, pipe cutter, hacksaw, muffler cutter, or chain cutter to cut the old system at convenient points to make the exhaust assembly smaller.

Although most exhaust systems use a slip joint and clamps to fasten the pipe to the muffler, a few use a welded connection. If the vehicle's system is welded, cut the pipe at the joint with a hacksaw or pipe cutter. The new pipe need not be welded to the muffler.

An adapter, available with the pipe, can be used instead. When measuring the length for the new pipe, allow at least 2 inches for the adapter to enter the muffler. The old exhaust pipe might be rusted into the muffler or converter opening. Attempt to collapse the old pipe by using a cold chisel or slitting tool and a hammer. While freeing the pipe, try not to damage the muffler inlet. It must be perfectly round to accept the new pipe.

Slide the new pipe into the muffler (some lubricant might be helpful). Attach the front end to the manifold. The pipe must fit at least 1-1⁄2 inches into the converter or muffler. A new gasket must be used at the manifold. Before tightening the connectors, check the system for alignment. When it is properly aligned, tighten the clamps.

If the vehicle is equipped with a mechanical heat riser valve, it should be sprayed with graphite oil. Also be sure that the counterweight is in balance. To do this, use pliers to move the counterweight and valve shaft slowly and carefully. Graphite oil is applied liberally during this process until the shaft moves freely and easily.

When installing exhaust gaskets, carefully follow the recommendations on the gasket package label and instruction forms. Read through all installation steps before beginning. Take note of any of the original equipment manufacturer's recommendations in service manuals that could affect engine sealing. Manifolds warp more easily if an attempt is made to remove them while they are still hot. Remember, heat expands metal, making assembly bolts more difficult to remove and easier to break.

To replace an exhaust manifold gasket, follow the torque sequence in reverse to loosen each bolt. Repeat the process to remove the bolts. This minimizes the chance that components will warp.

Any debris left on the sealing surfaces increases the chance of leaks. A good gasket remover will quickly soften the old gasket debris and adhesive for quick removal. Carefully remove the softened pieces with a scraper and a wire brush. Be sure to use a nonmetallic scraper when attempting to remove gasket material from aluminum surfaces.

Inspect the manifold for irregularities that might cause leaks, such as gouges, scratches, or cracks. Replace it if it is cracked or badly warped. File down any imperfections to ensure proper sealing of the manifold.

Due to high heat conditions, it is important to re-tap and rethread all threaded bolt holes, studs, and mounting bolts. This procedure ensures tight, balanced clamping forces on the gasket. Lubricate the threads with a good high-temperature antiseize lubricant. Use a small amount of contact adhesive to hold the gasket in place. Align the gasket properly before the adhesive dries. Allow the adhesive to dry completely before proceeding with manifold installation.

Install the bolts finger-tight. Tighten the bolts in three steps, one-half, three-quarters, and full torque, following the torque tables in the service manual or gasket manufacturer's instructions. When installing manifolds, the torquing of the bolts is usually begun in the center of the manifold, working outward in an X pattern. Be sure no exhaust part comes into direct contact with any section of the body, fuel lines, fuel tank, or brake lines.

Task D.4 Install appropriate exhaust system hardware (clamps, hangers, gaskets, flanges, fasteners, and heat shields).

Clamps, brackets, and hangers are used to properly join and support the various parts of the exhaust system. These parts also help to isolate exhaust noise by preventing its transfer through the frame or body to the passenger compartment. Clamps help to secure exhaust system parts to one another. The pipes are formed in such a way that one slips inside the other. This design makes a close fit. A U-type clamp usually holds this connection tight. Another important job of clamps and brackets is to hold pipes to the bottom of the vehicle. Clamps and brackets must be designed to allow the exhaust system to vibrate without transferring the vibrations through the car.

U-bolts of the same size as the exhaust pipe's outside diameter are used to clamp together sections of pipe. One end of one of the pipes is expanded to fit over the other piece of pipe. Clamps provide leak-free connections at joints in the muffler system.

Gaskets, usually made from asbestos, pressed steel, or sintered iron, are used to ensure tighter connections between the exhaust manifold and the exhaust pipe. The thick pipe ends on the converter and elsewhere require clamps that have to withstand anywhere from 30 to 60 ft.-lb. of torque. If the original clamp is a high-strength type with a welded saddle and hardened nuts, use an OE or OE-equivalent clamp for replacement.

Some exhaust systems are a single unit in which the pieces are welded together by the factory. By welding instead of clamping the assembly together, car makers save the weight of overlapping joints as well as that of clamps.

There are many different types of hangers available. Each is designed for a particular application. Some exhaust systems are supported by doughnut-shaped rubber rings between hooks on the exhaust component and on the frame or car body. Others are supported at the exhaust pipe and tailpipe connections by a combination of metal and reinforced fabric hanger. Both the doughnuts and the reinforced fabric allow the exhaust system to vibrate without breakage that could be caused by direct physical connection to the vehicle's frame.

Hangers hold the clamps, pipes, and muffler to the underside of the car. They are flexible to absorb road vibration; though if they are too flexible, road vibration can break them. On the other hand, if they are too firm, vibration and noise from the exhaust system are transferred to the underbody of the car and can be heard in the passenger compartment.

A hanger is usually a piece of fabric and rubber that resembles a tire. Each end is riveted to a piece of metal that is fastened on one end to the car body or frame. The other end is clamped to the pipe or muffler. The hanger allows the exhaust to be positioned away from other parts of the car and allows some flexibility from engine torque and vibration. Using a piece of rubber isolates the exhaust noise from the car body.

Hangers should be of the proper strength and design, as specified by the original manufacturer. They must also be strong enough to hold the weight of the system while allowing for flex. Also, make sure the correct number of hangers is being used.

Heat shields are located in areas where the exhaust system components (especially the catalytic converter and muffler) are close to the car body or near the ground. The shields reduce the amount of heat transferred into the car's body and protect items under the vehicle. They are usually made of pressed or perforated sheet metal. Heat shields trap the heat in the exhaust system, which has a direct effect on maintaining exhaust gas velocity.

Always reinstall all exhaust system heat shields. If the heat shields are not installed, undercoating, carpeting, dry leaves on the ground, and other flammable materials could catch on fire!

Task D.5 Inspect system for proper exhaust components clearance and routing.

Exhaust systems are designed for the vehicle they are in and the engine that is in the vehicle. In-line engines have a simple design that runs down one side of the engine. Called a single-exhaust system, there is only one pipe to the rear of the car. Single exhausts can be found on all engine sizes. A single-exhaust system has one path for exhaust flow through the system. Typically, it has only one header pipe, a main catalytic converter, a muffler, and a tailpipe. This is the most common type of exhaust system.

V-type engines with dual exhaust have two pipes, two mufflers, and two catalytic converters. A dual exhaust allows better breathing when the engine is under load. A dual-exhaust system has two separate exhaust paths to reduce backpressure. A crossover pipe joins the two manifolds into one pipe as they leave the engine on a V-type engine's single-exhaust system. A crossover pipe normally connects the right and left header pipes to equalize backpressure in a dual system. This also increases engine power slightly.

Though most cars and light trucks have exhaust systems with only a single catalytic converter, there are some with dual converters or piggyback converters. Service and repair procedures are pretty much the same on the dual-converter systems as those with only a single converter.

Unless exact duplicates of the original exhaust system are used, the system routing may not be correct. Always make sure the installed system does not come into contact with body, suspension, drivetrain, brake, or fuel system parts. Also, be sure no exhaust part comes into direct contact with any section of the body, fuel lines, fuel tank, or brake lines.

After the system is installed, check the clearances by moving the components against the hangers. Make sure no component can contact the underbody or frame of the vehicle or any other part. Remember the system must be able to move. If the system is rigidly mounted, exhaust vibrations will carry into the passenger compartment and may break prematurely.

Task D.6 Inspect system for proper exhaust component-to-component connection sealing.

In order for exhaust components to be completely sealed to one another, all pipes and pipe connections must be round. Leaks can also result from improperly positioned, under-tightened, and over-tightened clamps. Always use new gaskets when assembling the system. Be sure no exhaust part comes into direct contact with any section of the body, fuel lines, fuel tank, or brake lines.

One way of checking for leaks in a new system besides just listening is probing the system with an exhaust analyzer. With the engine running, slowly move the probe of the gas analyzer along the entire exhaust system. Before doing this, check the readings of the air in the shop. This is the base to compare against while you are probing. If, while moving the probe across the system, there is an increase in the readings, there is a leak. The key to finding the leak is remembering that it takes the analyzer approximately seven seconds to process a reading from the probe. Therefore if the readings increase, the leak is in an area you checked seven seconds ago. If you suspect an area as leaking, move the probe away from the exhaust and allow the readings to return to base. Then hold the probe at the suspected spot. If the readings begin to rise, you have found the leak. Make sure you inspect all exhaust system joints for leaks. These leaks must be corrected to protect the vehicle's occupants from being exposed to toxic gases.

Task D.7 Install exhaust subsystem components [air injection reactor (AIR), exhaust gas recirculation (EGR) valve, oxygen sensor(s) (O_2S/HO_2S), early fuel evaporation (EFE) system (heat riser)].

As part of servicing automotive exhaust systems, a technician must be able to service and install exhaust and emission control subsystem components. These include air injection reactor (AIR), exhaust gas recirculation (EGR) valve, oxygen sensor(s) both (O_2S/HO_2S), and early fuel evaporation (EFE) system components. It is extremely important that these subsystem components are not disabled, tampered with, or rendered inoperative during service or repair, as they are an integral part of the exhaust system design and function.

E. Exhaust System Repair Regulations (7 Questions)

Task E.1 Comply with warranty and diagnostic requirements regarding permissible catalytic converter installations.

The EPA has established regulations, at the federal level, concerning all emission control devices. Individual states, counties, and cities have the right to institute their own regulations providing these meet the federal standards. Always make sure you know the laws that directly affect you and your occupation. The following describes the federal standards.

Replacement converters must have documents or labels that show they meet EPA requirements and are warranted to meet federal durability and performance standards. All manufacturers of new and rebuilt converters who meet the EPA requirements must state that fact in writing, usually in the warranty information in their catalogs. If you bypass a damaged or ineffective catalytic converter instead of replacing it or repairing it, you are breaking the law! The penalty could be much higher than the price of a new or rebuilt unit.

Federally required emission control warranties protect the vehicle's owner from the cost of repairs for certain emission-related failures that result from manufacturer defects in materials and workmanship or that cause the vehicle to exceed federal emission standards. Manufacturers have been required by federal law to provide emission control coverage for vehicles since 1972. There are two federal emission control warranties discussed here: (1) performance warranty and (2) design and defect warranty.

The performance warranty covers repairs which are required during the first 2 years of 24,000 miles of vehicle use because the vehicle failed an emission test. Specified major emission control components are covered for the first 8 years or 80,000 miles. If the owner of the vehicle is a resident of an area with an inspection and maintenance (I/M) program that meets federal guidelines, the owner is eligible for this warranty protection provided that: (1) the car or light-duty truck fails an approved emissions test; (2) the vehicle is less than 2 years old and has less than 24,000 miles (up to 8 years/ 80,000 miles for certain components); (3) the state or local government requires that the vehicle be repaired; (4) the vehicle's test failure does not result from misuse of the vehicle or a failure to follow the manufacturer's written maintenance instructions; and (5) the owner presented the vehicle to a warranty-authorized manufacturer representative, along with evidence of the emission test failure, during the warranty period.

During the first 2 years/24,000 miles, the performance warranty covers any repair or adjustment which is necessary to make the vehicle pass an approved, locally required emission test and as long as the vehicle has not exceeded the warranty time or mileage limitations and has been properly maintained according to the manufacturer's specifications.

The design and defect warranty covers repair of emission-related parts, which become defective during the warranty period. The design and defect warranty for model year 1995 and newer light-duty cars and trucks covers emission control and emission-related parts for the first 2 years or 24,000 miles of vehicle use and specified major emission control components for the first 8 years or 80,000 miles of vehicle use.

According to federal law, an emission control or emission-related part, or a specified major emission control component, that fails because of a defect in materials or workmanship, must be repaired or replaced by the vehicle manufacturer free of charge as long as the vehicle has not exceeded the warranty time or mileage limitations for the failed part.

An emission control part is any part installed with the primary purpose of controlling emissions. An emission-related part is any part that has an effect on emissions. Some examples of parts or systems that fall under these definitions follow. A more complete list can be found in the owner's manual/warranty booklet. If any of these parts fail to function or function improperly because of a defect in materials or workmanship, causing the vehicle to exceed federal emission standards, they should be repaired or replaced under the emissions warranty if the vehicle is less than 2 years old and has been driven less than 24,000 miles. One manufacturer may use more parts than another, so this list is not complete for all vehicles. Parts with a stated replacement interval, such as, "replace at 15,000 miles or 12 months," are warranted up to the first replacement point only.

There are three specified major emission control components covered for the first 8 years or 80,000 miles of vehicle use on 1995 and newer vehicles: catalytic converters, the electronic emissions control unit or computer (ECU), and the on-board emissions diagnostic device or computer (OBD).

Catalytic converters are critical emission control components that have been installed on most cars and trucks manufactured since 1975. Since engines do not burn fuel completely during the combustion process, the exhaust contains a significant amount of harmful pollutants such as carbon monoxide, hydrocarbons, and oxides of nitrogen. The catalytic converter aids the conversion of these pollutants to less harmful substances such as carbon dioxide, water vapor, nitrogen, and oxygen before the exhaust is expelled into the environment.

If a qualified automotive technician or owner can show that an emission control or emission-related component, or a specified major, emission control component, is defective, the repair or replacement of the part is probably covered under the design and defect warranty. If the vehicle failed a federally approved emissions test and has not exceeded the time and mileage limitations for the performance warranty, any repairs or adjustments necessary for the vehicle to pass should be covered by the manufacturer if the failure was not caused by improper maintenance or abuse. If the vehicle has a defective part, or the vehicle fails an emission test, the procedures for making a warranty claim as identified by the manufacturer in the owner's manual or warranty booklet should be followed. To process the warranty, a copy of the I/M test report as proof of the emissions test failure must be submitted.

If the part or system should be covered by the warranties, the owner must take the vehicle to a facility authorized by the vehicle manufacturer for repair to give them the opportunity to diagnose and repair it. Note that if your facility is not authorized by the vehicle manufacturer, you are not obligated to advise the owner of the parts that are covered under warranty.

The emissions warranties apply to used vehicles, as well as new ones, as long as the vehicle has not exceeded the warranty time or mileage limitations.

If the vehicle has had all the proper maintenance performed and it still fails an emissions test, the catalytic converter should be checked. Occasionally, the converter may need to be replaced so a vehicle can pass the test. Before undertaking this major repair, be sure to do a thorough diagnosis of the engine and emission system. Catalytic converters rarely fail without a cause. A problem in the catalyst is usually the result of a problem upstream in the engine management system. If it must be replaced because it is missing or is no longer functioning correctly, make certain the replacement's label shows that it meets EPA requirements and is warranted to meet federal durability and performance standards. All manufacturers of new and rebuilt converters who meet the EPA requirements must state that fact in writing. If you bypass a damaged or ineffective catalytic converter instead of replacing it or repairing it, you are breaking the law. The penalty could be much higher than the price of a new or rebuilt unit.

EPA regulations require that 1994 model year cars and light trucks have on-board diagnostic computers that monitor the operation of emissions control systems. Information on malfunctions is stored in the system's memory so that a technician can retrieve them. Because many problems with the emissions control system do not affect a vehicle's performance, drivers may not be aware of the problem (unless, of course, the vehicle fails an emissions test). When the computer identifies a problem, it will alert the driver via a dashboard light.

Task E.2 Comply with requirements regarding prohibited catalytic converter installations.

The EPA has established regulations, at the federal level, concerning all emission control devices. Individual states, counties, and cities have the right to institute their own regulations providing these meet the federal standards. Always make sure you know the laws that directly affect you and your occupation. The following describes the federal standards.

Replacement converters must have documents or labels that show they meet EPA requirements and are warranted to meet federal durability and performance standards. All manufacturers of new and rebuilt converters who meet the EPA requirements must state that fact in writing, usually in the warranty information in their catalogs. If you bypass a damaged or ineffective catalytic converter instead of replacing it or repairing it, you are breaking the law! The penalty could be much higher than the price of a new or rebuilt unit.

If the vehicle is within the age and mileage limits for the applicable emissions warranty, the manufacturer can only deny coverage if evidence shows that the owner has failed to properly maintain and use the vehicle, causing the part or emission test failure.

Some examples of misuse and lack of maintenance include the following: vehicle abuse such as off-road driving or overloading; tampering with emission control parts or systems, including removal or intentional damage of such parts or systems; and improper maintenance, including failure to follow maintenance schedules and instructions specified by manufacturer, or use of replacement parts which are not equivalent to the originally installed parts.

Replacement catalytic converters must not only meet EPA standards but they must also be the exact type and be an exact replacement for the original converter.

Task E.3 Comply with requirements regarding record keeping.

The converter is one of the components covered by federal emission regulations, so it cannot be replaced by anyone other than a car dealer unless it is out of warranty and a legitimate need for replacement (such as failing an emissions test, detecting damage or blockage, or a missing converter) has been established and documented. Up to model year 1995, converters were covered by a 5-year/50,000-mile federal emissions warranty (7 years or 70,000 miles in California). In 1995, the warranty jumped to 8 years and 80,000 miles. You must also obtain the customer's authorization for repairs in writing; keep the paperwork for six months and the old converter for 15 days. The replacement converter must be the same type as the original and be installed in the same location.

Replacement converters must have documents or labels that show they meet EPA requirements and are warranted to meet federal durability and performance standards. All manufacturers of new and rebuilt converters who meet the EPA requirements must state that fact in writing, usually in the warranty information in their catalogs. If you bypass a damaged or ineffective catalytic converter instead of replacing it or repairing it, you are breaking the law! The penalty could be much higher than the price of a new or rebuilt unit.

Task E.4 Comply with requirements regarding catalytic converter replacement, location, and type.

The EPA has established regulations, at the federal level, concerning all emission control devices. Individual states, counties, and cities have the right to institute their own regulations providing these meet the federal standards. Always make sure you know the laws that directly affect you and your occupation. The following describes the federal standards.

Replacement converters must have documents or labels that show they meet EPA requirements and are warranted to meet federal durability and performance standards. All manufacturers of new and rebuilt converters who meet the EPA requirements must state that fact in writing, usually in the warranty information in their catalogs. If you bypass a damaged or ineffective catalytic converter instead of replacing it or repairing it, you are breaking the law! The penalty could be much higher than the price of a new or rebuilt unit.

If the vehicle is within the age and mileage limits for the applicable emissions warranty, the manufacturer can only deny coverage if evidence shows that the owner has failed to properly maintain and use the vehicle, causing the part or emission test failure.

Some examples of misuse and lack of proper maintenance include the following: vehicle abuse such as off-road driving or overloading; tampering with emission control parts or systems, including removal or intentional damage of such parts or systems; and improper maintenance, including failure to follow maintenance schedules and instructions specified by manufacturer, or use of replacement parts which are not equivalent to the originally installed parts.

Replacement catalytic converters must not only meet EPA standards, but they must also be the exact type and an exact replacement for the original converter. They must also be placed in the exact same location as the original.

Task E.5 Comply with requirements regarding replacement of air injection reactor (AIR), exhaust gas recirculation (EGR), oxygen sensor (O_2S/HO_2S), heat riser [early fuel evaporation (EFE)], and turbocharger systems.

Emission controls on cars and trucks have one purpose: to reduce the amount of pollutants and environmentally damaging substances released by the vehicles. The consequences of the pollutants are grievous. The air we breathe and the water we drink have become contaminated with chemicals that adversely affect our health. It took many years for the public and the industry to address the problem of these pollutants. Not until smog became an issue did anyone in power really care and do something about these pollutants.

Smog not only appears as dirty air, it is also an irritant to your eyes, nose, and throat. The things necessary to form photochemical smog are HC and NO_x exposed to sunlight in stagnant air. When there is enough HC in the air, it reacts with the NO_x in the air. The energy of sunlight causes these two chemicals to react and form photochemical smog.

There are three main automotive pollutants: hydrocarbons (HC), carbon monoxide (CO), and oxides of nitrogen (NO_x). Particulate emissions are also present in diesel engine exhaust. HC emissions are caused largely by unburned fuel from the combustion chambers. HC emissions can also originate from evaporative sources such as the gasoline tank. CO emissions are a by-product of the combustion process, resulting from incorrect air-fuel mixtures. NOx emissions are caused by nitrogen and oxygen uniting at cylinder temperatures above 2,500°F (1,371°C).

The first Clean Air Act prompted Californians to create the California Air Research Board. California ARB's purpose was to implement strict air standards; these became the standard for federal mandates. One of the approaches to clean the air by the ARB was to start periodic motor vehicle inspection (PMVI). The purpose for the PMVI is to inspect a vehicle's emission controls once a year. This inspection included a tailpipe emissions test and an under-hood inspection. The tailpipe test certifies that the vehicle's exhaust emissions are within the limits set by law. The underhood and/or vehicle inspection verifies that the pollution control equipment has not been tampered with or disconnected.

Today, California is not the only state that requires annual emissions testing. Many states have incorporated an emissions test with their annual vehicle registration procedures. Most states have or are planning to implement an I/M 240 or similar program. The I/M 240 tests the emissions of a vehicle while it is operating under a variety of load conditions and speeds. This is an improvement over exhaust testing during idle and high speed with no load.

The I/M 240 test requires the use of a chassis (road) dynanameter, commonly called a dyno. While on the dyno, the vehicle is operated for 240 seconds and under different load conditions. The test drive on the dyno simulates both in-traffic and highway driving and stopping. The emissions tester tracks the exhaust quality through these conditions.

The I/M 240 program also includes a functional test of the evaporative emission control devices and a visual inspection of the total emission control system. If the vehicle fails the test, it must be repaired and certified before it can be registered.

According to a document based on a study by the Environmental Protection Agency (EPA), passenger cars are responsible for 17.8 percent of the total hydrocarbon emissions, 30.9 percent of the total carbon monoxide emissions, and 11.1 percent of the oxides of nitrogen emissions. After more than 30 years of emission regulations, these figures remain staggering! Imagine what these figures would be if automotive and industrial emissions had remained unregulated during the last 30 years!

Emission standards have been one of the driving forces behind many of the technological changes in the automotive industry. Catalytic converters and other emission systems were installed to meet emission standards. Computer-controlled carburetors and fuel injection systems were installed to provide more accurate control of the air-fuel ratio to reduce emission levels and allow the catalytic converter to operate efficiently.

During the 1990s, emission standards in the United States became increasingly stringent. In 1994, an ambitious emission program began in California. This California program specifies emission standards for transitional low emission vehicles (TLEV). In 1997, the California emission program specified a further reduction in emission levels for low emission vehicles (LEV), and in the year 2000, emission levels are specified for ultra low emission vehicles (ULEV).

In 1998, California emission regulations required 2 percent of the cars sold in the state to be zero emission vehicles (ZEV). With present technology, only electric cars will meet the ZEV standards. Some other states have adopted, or are considering the adoption of, the California emission standards. The percentage of ZEVs increases in 2003 and 2005.

Present government goals call for a 98 percent reduction of unburned hydrocarbons, a 97 percent reduction of carbon monoxide, and a 90 percent reduction of oxides of nitrogen compared to pre-controlled cars.

Many changes will be necessary in vehicles, engines, and emission systems to meet the emission standards of the twenty-first century. Since approximately 90 percent of hydrocarbon emissions occur before the catalytic converter is hot enough to provide proper HC oxidation, engineers are designing catalytic converters that are heated from the vehicle electrical system. Since these heaters have high current flow requirements, some charging system modifications may be necessary. This is just one of the many changes in emission systems that we will likely see in the near future.

The federal government has set standards for these pollutants. The exceptions to these standards are a few high-altitude western states and California. Because there is less oxygen at high altitudes to promote combustion, emission standards at high altitudes are slightly less strict. Because of the dense population of the state, California's standards allow less pollution than federal standards. Many states have followed California's mandate and have instituted emission standards of the same levels as California.

Automobile manufacturers have been working toward reduction of automotive air pollutants since the early 1950s, when auto emissions were found to be part of the cause of smog in Los Angeles. Governmental interest in controlling emissions developed around the same time.

The Environmental Protection Agency establishes emissions standards that limit the amount of these pollutants a vehicle can emit. To meet these standards, many changes have been made to the engine itself. Plus there have been systems developed and added to the engines to reduce the pollutants they emit. A list of the most common pollution control devices related to the exhaust system follows.

- *Exhaust gas recirculation (EGR) system.* Introduces exhaust gases into the intake air to reduce the temperatures reached during combustion. This reduces the chances of forming NO_x during combustion.
- *Catalytic converter.* Located in the exhaust system, it allows for the burning or converting of HC, CO, and NO_x into harmless substances, such as water.
- *Air-injection system.* This system reduces HC emissions by introducing fresh air into the exhaust stream to cause minor combustion of the HC in the engine's exhaust.

Three basic types of emission control systems are used in modern vehicles: evaporative control systems, pre-combustion, and post-combustion. The evaporative control system is a sealed system. It traps the fuel vapors (HC) that would normally escape from the fuel tank and carburetor into the air.

Most of the pollution control systems used today prevent emissions from being created in the engine, either during or before the combustion cycle. The common pre-combustion control systems are the positive crankcase ventilation (PCV), engine modification systems, and exhaust gas recirculating (EGR) systems.

Post-combustion control systems clean up the exhaust gases after the fuel has been burned. Secondary air or air injector systems put fresh air into the exhaust to reduce HC and CO to harmless water vapor and carbon dioxide by chemical (thermal) reaction with oxygen in the air. Catalytic converters help this process. Most catalysts reduce NO_x as well as HC-CO.

OBD-II systems must illuminate the MIL if the vehicle conditions would allow emissions to exceed 1.5 times the allowable standard for that model year based on a federal test procedure (FTP). When a component or strategy failure allows emissions to exceed this level, the MIL is illuminated to inform the driver of a problem and a diagnostic trouble code is stored in the PCM.

Besides enhancements to the computer's capacities, some additional hardware is required to monitor the emissions performance closely enough to fulfill the tighter constraints and beyond merely keeping track of component failures. In most cases, this hardware consists of an additional heated oxygen sensor down the exhaust stream from the catalytic converter, upgrading specific connectors and components to last the mandated 100,000 miles or 10 years, in some cases a more precise crankshaft or camshaft position sensor (to detect misfires), and a new standardized 16-pin DLC.

The California Air Resources Board (CARB) found that by the time a computer or emission system component failure occurs and the malfunction indicator light is illuminated, the vehicle emissions have been excessive for some time. The CARB developed requirements to monitor the performance of emission systems, as well as to indicate component failure. These requirements were accepted by the EPA. The monitoring results must be available to service personnel without special test equipment marketed by the vehicle manufacturer. The monitoring system for engine, computer system, and emission system equipment is called OBD-II.

Computer systems without OBD-II have the ability to detect component and system failure. Computer systems with OBD-II are capable of monitoring the ability of systems and components to maintain low emission levels.

Computer systems with OBD-II capabilities are similar to previous systems except for the monitoring systems and the monitoring strategies in the PCM, which are extensive. New refinements are frequently incorporated into the PCM and other system components as improved technology is developed. Monitors included in OBD-II are:

1. Catalyst efficiency monitor
2. Engine misfire monitor
3. Fuel system monitor
4. Heated exhaust gas oxygen sensor monitor
5. Exhaust gas recirculation monitor
6. Evaporative system monitor
7. Secondary air-injection monitor
8. Comprehensive component monitor

It is against the law for anyone to tamper, remove, or intentionally damage any part of the emission control system. When inspecting an exhaust system, carefully look at the system and attempt to identify any signs of tampering. Sometimes the law violations are obvious, such as the presence of a straight pipe where the catalytic converter should be. Some tampering is not as quickly seen, such as the enlargement of the fuel filler neck so that it can accept leaded fuels. Use of leaded fuels is deadly to the catalytic converter. Also check the piping to the AIR and EGR systems. Any tampering must be noted prior to doing any repair work and must be corrected during the repair procedure. Failure to follow these guidelines would make you liable for any and all fines levied against the owner of the vehicle.

If a catalytic converter is found to be bad, under no circumstances should the converter be removed and replaced with a straight piece of pipe (a test pipe). This practice is illegal for the professional auto technician. Because of constant change in EPA catalytic converter removal and installation requirements, check with the manufacturer or EPA for the latest data regarding replacement.

Replacement converters must have documents or labels that show they meet EPA requirements and are warranted to meet federal durability and performance standards. All manufacturers of new and rebuilt converters who meet the EPA requirements must state that fact in writing, usually in the warranty information in their catalogs. If you bypass a damaged or ineffective catalytic converter instead of replacing it or repairing it, you are breaking the law! The penalty could be much higher than the price of a new or rebuilt unit.

Task E.6 Comply with requirements regarding exhaust system configuration.

It is against the law for anyone to tamper, remove, or intentionally damage any part of the emission control system. When inspecting an exhaust system, carefully look at the system and attempt to identify any signs of tampering. Sometimes the law violations are obvious, such as the presence of a straight pipe where the catalytic converter should be. Some tampering is not as quickly seen, such as the enlargement of the fuel filler neck so that it can accept leaded fuels. Use of leaded fuels is deadly to the catalytic converter.

Also check the piping to the AIR and EGR systems. Any tampering must be noted prior to doing any repair work and must be corrected during the repair procedure. Failure to follow these guidelines would make you liable for any and all fines levied against the owner of the vehicle.

There may be strict regulations preventing or defining exhaust configurations. This may include the use of anything but exact replacement parts or altering the flow paths. Always be aware of the local laws before changing anything in an exhaust system.

If a catalytic converter is found to be bad, under no circumstances should the converter be removed and replaced with a straight piece of pipe (a test pipe). This practice is illegal for the professional auto technician. Because of constant change in EPA catalytic converter removal and installation requirements, check with the manufacturer or EPA for the latest data regarding replacement.

Replacement converters must have documents or labels that show they meet EPA requirements and are warranted to meet federal durability and performance standards. All manufacturers of new and rebuilt converters who meet the EPA requirements must state that fact in writing, usually in the warranty information in their catalogs. If you bypass a damaged or ineffective catalytic converter instead of replacing it or repairing it, you are breaking the law! The penalty could be much higher than the price of a new or rebuilt unit.

5 Sample Test for Practice

Sample Test

Please note the letter and number in parentheses following each question. They match the task in Section 4 that discusses the relevant subject matter. You may want to refer to the overview using the cross-referencing key to help with questions posing problems for you.

1. While trying to identify the cause of an exhaust noise, Undercar Specialist A gently taps against the pipes and muffler with a hammer or mallet and says a weak or worn-out part will have a dull sound. Undercar Specialist B grabs the tailpipe and tries to move it up and down and from side to side and says there should be only slight movement in any direction. Who is right?
   A. A only
   B. B only
   C. Both A and B
   D. Neither A nor B (A.1.1)

2. While cutting a pipe with an oxygen/acetylene torch, Undercar Specialist A places the tip of the inner cone of the preheat flame to the edge of the metal to be cut so that it is about one-half of an inch away. Undercar Specialist B squeezes the cutting oxygen lever as soon as a spot on the cutting line has been heated to a bright cherry red color. Who is right?
   A. A only
   B. B only
   C. Both A and B
   D. Neither A nor B (C.2.3)

3. Undercar Specialist A says that a catalytic converter breaks down HC and CO to relatively harmless by-products. Undercar Specialist B says that using leaded gasoline or allowing the converter to overheat can destroy its usefulness. Who is right?
   A. A only
   B. B only
   C. Both A and B
   D. Neither A nor B (B.1)

4. All of the following EGR valve problems can be the cause of spark knock **EXCEPT:**
   A. a valve stuck closed.
   B. a leaking valve diaphragm.
   C. restrictions in the exhaust flow passages.
   D. a stuck open valve. (B.3)

5. Undercar Specialist A says a single-exhaust system has only one exhaust path to the rear of the car. Undercar Specialist B says dual-exhaust systems start as a single system and split into two paths, each with a converter and muffler. Who is right?
   A. A only
   B. B only
   C. Both A and B
   D. Neither A nor B (D.1)

6. While conducting a visual inspection of an exhaust system, Undercar Specialist A uses a droplight and closely inspects the system for leaks and says soot or discoloration indicates escaping hot gases. Undercar Specialist B says overheated converters will have a brownish or bluish tint to them and it is normal for the area around the converter to be blistered because of the converter's operating temperature. Who is right?
   A. A only
   B. B only
   C. Both A and B
   D. Neither A nor B (A.1.1)

7. Which of the following is a likely cause for an OBD-II AIR monitor failure?
   A. Faulty EGR valve
   B. Disconnected or damaged AIR hoses and/or tubes
   C. An open H_2OS circuit
   D. A faulty PCV check valve (B.2)

8. After installing a new exhaust manifold on a vehicle with downstream AIR, the vacuum line to the air-switching valve was not connected properly. Technician A says that this could cause the vehicle to run rich. Technician B says that this could cause the catalytic converter to overheat. Who is right?
   A. A only
   B. B only
   C. Both A and B
   D. Neither A nor B (D.7)

9. While discussing EGR valve diagnosis, Undercar Specialist A says a defective throttle position sensor may affect the EGR valve operation. Undercar Specialist B says a defective engine coolant temperature (ECT) sensor may affect the EGR valve operation. Who is right?
   A. A only
   B. B only
   C. Both A and B
   D. Neither A nor B (B.3)

10. A catalytic converter case has a bluish color. Undercar Specialist A says the air-fuel ratio may be too rich. Undercar Specialist B says an ignition misfire could be the cause. Who is right?
   A. A only
   B. B only
   C. Both A and B
   D. Neither A nor B (A.1.2)

11. Before replacing any exhaust system component, Undercar Specialist A soaks all old connections with penetrating oil. Undercar Specialist B checks the old system's routing for critical clearance points. Who is right?
   A. A only
   B. B only
   C. Both A and B
   D. Neither A nor B (A.2.4)

12. When diagnosing a catalytic converter with the engine running at normal operating temperature, the converter:
    A. inlet should be 150°F (65°C) hotter than the outlet.
    B. outlet should be 100°F (38°C) hotter than the inlet.
    C. outlet should be 20°F (6.6°C) cooler than the inlet.
    D. inlet should be 50°F (19°C) hotter than the outlet. (B.1)

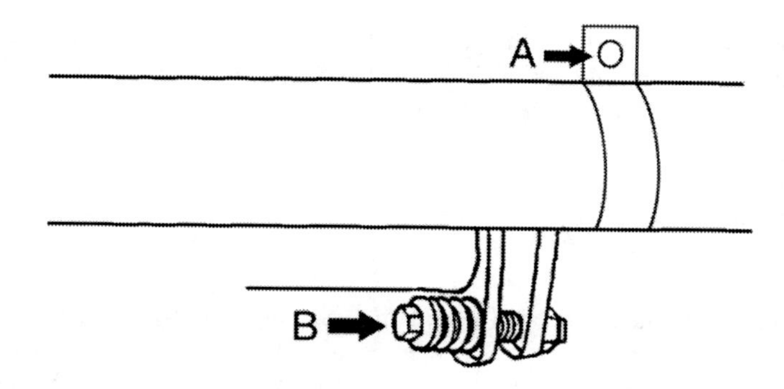

13. Undercar Specialist A says part A in the figure is a typical clamp that is used to join two exhaust parts together. Undercar Specialist B says part B is a spring-loaded hanger designed to minimize vibrations. Who is right?
    A. A only
    B. B only
    C. Both A and B
    D. Neither A nor B (A.2.4)

14. Undercar Specialist A says the vacuum-actuated heat riser defaults to open position when vacuum is removed from actuator. Undercar Specialist B says that it will close when vacuum signal is removed. Who is right?
    A. A only
    B. B only
    C. Both A and B
    D. Neither A nor B (B.4)

15. Undercar Specialist A says boost pressure is regulated by the wastegate. Undercar Specialist B says that higher boost pressures can be obtained with lower octane fuels. Who is right?
    A. A only
    B. B only
    C. Both A and B
    D. Neither A nor B (A.1.2)

16. Before using a MIG welder, Undercar Specialist A selects the wire size. Undercar Specialist B sets the unit to the desired heat range. Who is correct?
    A. A only
    B. B only
    C. Both A and B
    D. Neither A nor B (C.2.1)

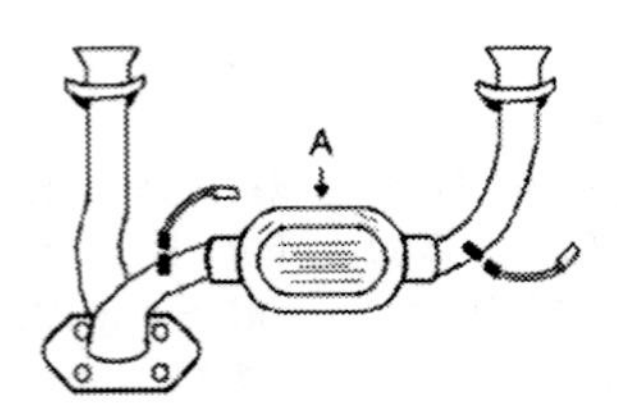

17. Undercar Specialist A says when the part shown in the figure is replaced, it is cut from the pipe and a new one welded into the unit. Undercar Specialist B says the part shown in the figure is a mini-converter in a crossover pipe. Who is right?
    A. A only
    B. B only
    C. Both A and B
    D. Neither A nor B (A.2.2)

18. Undercar Specialist A says if the voltage at an O_2S is continually high, the air-fuel ratio may be rich or the sensor may have a shorted heater circuit. Undercar Specialist B says when the O_2S voltage is continually low, the air-fuel ratio may be lean, the sensor may be defective, or the wire between the sensor and the computer may have a high-resistance problem. Who is right?
    A. A only
    B. B only
    C. Both A and B
    D. Neither A nor B (B.5)

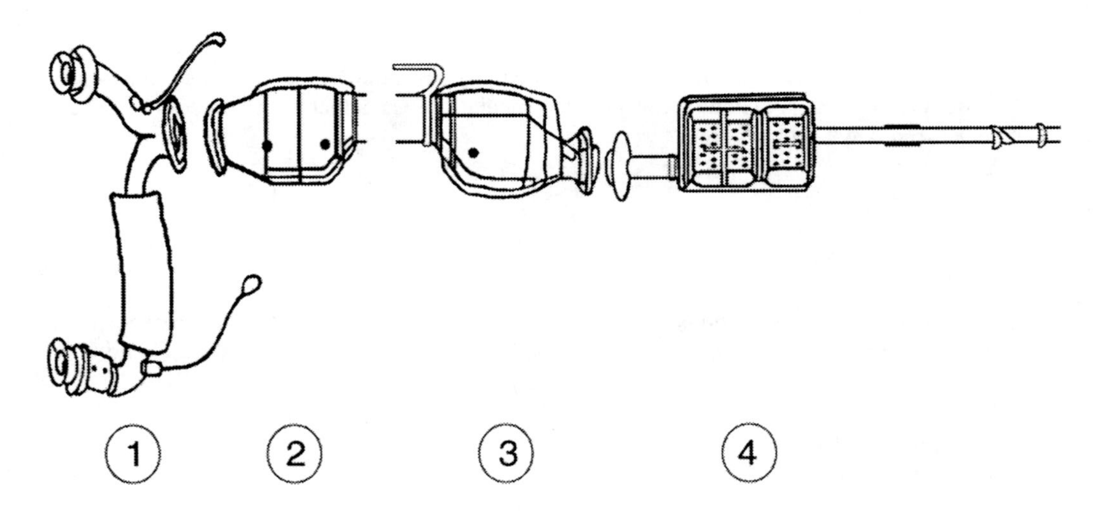

19. Part 2 in the figure is MOST Likely a:
    A. catalytic converter.
    B. muffler.
    C. resonator.
    D. mini-converter. (A.2.1)

20. Undercar Specialist A uses sandpaper to remove carbon deposits from turbocharger wastegate parts. Undercar Specialist B scrapes off heavy deposits before attempting to clean the unit. Who is right?
    A. A only
    B. B only
    C. Both A and B
    D. Neither A nor B
(A.1.3)

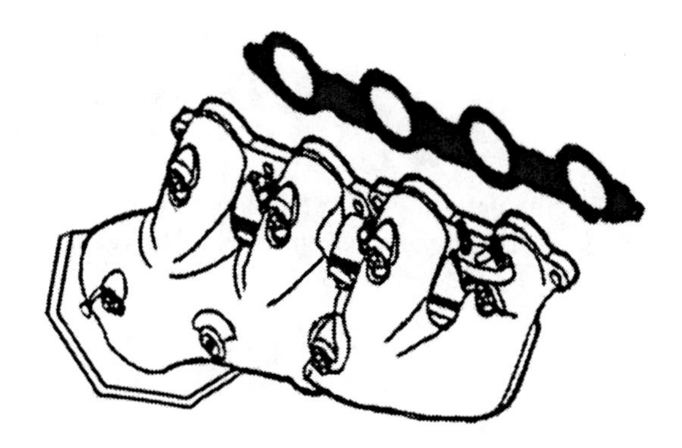

21. All of the following statements about the part shown in the figure are true **EXCEPT:**
    A. this part may warp because of excess heat.
    B. a straightedge and feeler gauge can be used to check for cracks in the machined surface of the part.
    C. heat may cause the part to crack.
    D. cracks in the part may occur when a car passes through a large puddle and cold water splashes on the hot surface.
(A.2.3)

22. Which of the following is NOT true about testing an oxygen sensor with a voltmeter?
    A. The signal from most oxygen sensors varies between 0 and 1 volt.
    B. If the oxygen sensor voltage signal remains in a mid-range position, the sensor is good.
    C. If the voltage is continually high, the air-fuel ratio may be rich or the sensor may be contaminated by RTV sealant, antifreeze, or lead from leaded gasoline.
    D. When the oxygen sensor voltage is continually low, the air-fuel ratio may be lean, the sensor may be defective, or the wire between the sensor and the computer may have a high-resistance problem.
(B.5)

23. Undercar Specialist A says when bending pipe for an exhaust system, the straight pipe should be inserted under the vehicle to determine the routing that would need a minimum amount of bends. Undercar Specialist B says some pipe-bending machines use program cards to bend the pipe at preset angles for the vehicle. Who is right?
    A. A only
    B. B only
    C. Both A and B
    D. Neither A nor B
(C.1.1)

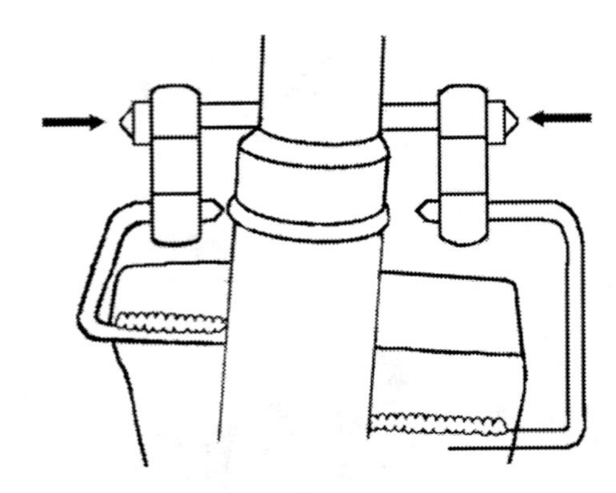

24. Undercar Specialist A says the parts shown in the figure keep the exhaust system in proper alignment. Undercar Specialist B says if these parts are damaged, the converter may tend to overheat. Who is right?
    A. A only
    B. B only
    C. Both A and B
    D. Neither A nor B (D.5)

25. While discussing emission systems that are related to the exhaust system, Undercar Specialist A says the EGR system introduces exhaust gases into the intake air to reduce the chances of forming NOx during combustion. Undercar Specialist B says the AIR reduces HC emissions by introducing fresh air into the exhaust stream to cause minor combustion of the HC in the engine's exhaust. Who is right?
    A. A only
    B. B only
    C. Both A and B
    D. Neither A nor B (A.1.3)

26. All of these statements are true **EXCEPT:**
    A. heat shields are used to protect the catalytic converter from the heat of the exhaust system.
    B. heat shields trap the heat in the exhaust system, which has a direct effect on maintaining exhaust gas velocity.
    C. clamps, brackets, and hangers are used to isolate exhaust noise by preventing its transfer through the frame or body to the passenger compartment.
    D. clamps help to secure the exhaust system parts to one another. (D.2)

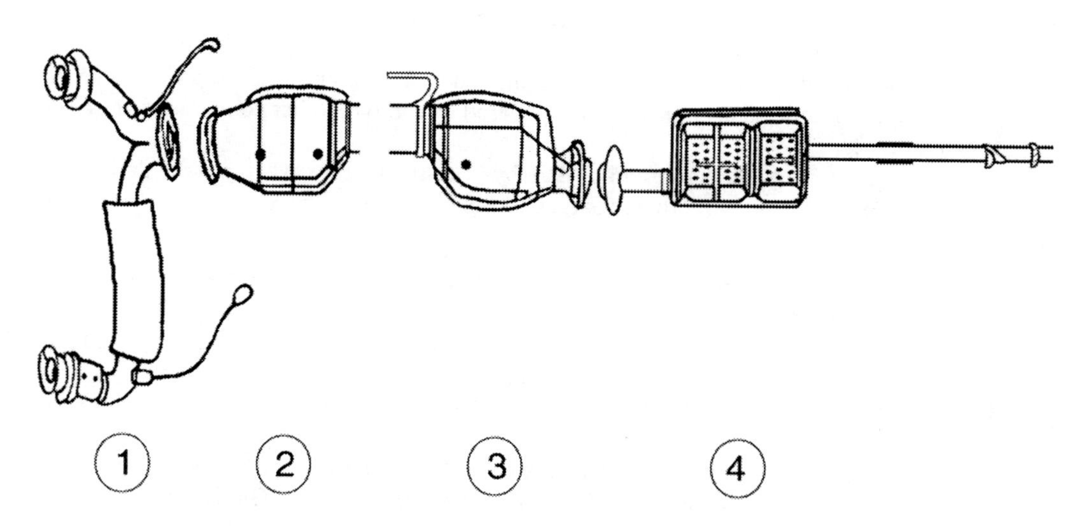

27. Part 4 in the figure is MOST Likely a:
    A. catalytic converter.
    B. muffler.
    C. resonator.
    D. mini-converter. (A.2.5)

28. Undercar Specialist A makes sure the depth of the bend is exactly what is required. Undercar Specialist B bends the pipe as required to fit it under the vehicle. Who is right?
    A. A only
    B. B only
    C. Both A and B
    D. Neither A nor B (C.1.2)

29. All of the following could cause a rough idle **EXCEPT:**
    A. restricted or clogged exhaust port in the EGR valve.
    B. a stuck open EGR valve.
    C. dirt on the EGR's valve seat.
    D. loose mounting bolts. (A.1.3)

30. While checking the seal of newly installed exhaust system, Undercar Specialist A carefully listens at each joint of the exhaust system. Undercar Specialist B probes each joint with an exhaust analyzer. Who is right?
    A. A only
    B. B only
    C. Both A and B
    D. Neither A nor B (D.6)

31. Undercar Specialist A says worn, damaged, or broken engine/transmission mounts could be the cause of exhaust system rattles. Undercar Specialist B says worn, damaged, or broken engine/transmission mounts could be the cause of exhaust system breakage. Who is right?
    A. A only
    B. B only
    C. Both A and B
    D. Neither A nor B (A.1.6)

32. Which of the following is NOT a true statement about fabricating an exhaust pipe?
    A. Once the pipe has all of its required bends, it is ready to be installed.
    B. The pipe must be cut to the correct length; make sure the cuts are straight and even.
    C. The ends of an exhaust pipe must be prepared to join with other components or to serve as a tailpipe.
    D. Check a newly made flange and/or flare by inserting a new gasket into the flare to make sure it is round and will provide a good sealing area for the gasket. (C.1.3)

33. Which of the following statements about EGR systems is NOT true?
    A. The EGR system dilutes the air-fuel mixture with controlled amounts of exhaust gas to reduce peak combustion temperatures.
    B. For drivability, it is desirable to have the EGR valve opening (and the amount of gas flow) proportional to the throttle opening.
    C. Drivability is improved by opening the EGR when the engine is started up cold, at idle, and at full throttle.
    D. Because NO_x control requirements vary on different engines and operating conditions, there are several different control systems used in the various EGR systems. (A.1.3)

34. Which of the following statements about exhaust pipes is NOT true?
    A. The exhaust pipe carries exhaust gases and vapor out into the air and directs them where they cannot enter the passenger compartment.
    B. A "Y" pipe is an exhaust pipe that connects both exhaust manifolds of a V-type engine to form a single-exhaust system.
    C. An "H" pipe consists of right and left exhaust pipes connected by a balance pipe that forms a dual-exhaust system.
    D. The intermediate pipe connects the exhaust pipe with the muffler or resonator and is also known as an extension or connecting pipe. (A.2.1)

35. All of the following statements about oxyacetylene welding are true **EXCEPT:**
    A. oxyacetylene welding is a manual process; the welder controls the torch movement and the welding rod.
    B. one torch handle valve controls the rate of oxygen flow and the other controls the rate of acetylene flow.
    C. the two gases mix and begin to burn in the handle of the torch.
    D. oxyacetylene welding joins metals by melting and fusing. (C.2.1)

36. Undercar Specialist A says when replacing the EGR valve, a new plate gasket should be installed. Undercar Specialist B says that if the plate gasket looks serviceable, it can be reused. Who is correct?
    A. A only
    B. B only
    C. Both A and B
    D. Neither A nor B (D.4)

37. All of the following are steps in an inspection of a turbocharger system, **EXCEPT:**
    A. check the air cleaner and remove the ducting from the air cleaner to turbo and look for dirt buildup or damage from foreign objects.
    B. check for loose clamps on the compressor outlet connections and check the engine intake system for loose bolts or leaking gaskets.
    C. disconnect the exhaust pipe and look for restrictions or loose material.
    D. remove the turbo shaft assembly and look for signs of rubbing or wheel impact damage. (A.1.3)

38. Undercar Specialist A says replacement catalytic converters must meet EPA standards. Undercar Specialist B says replacement converters are available in two different sizes and undercar specialists should choose the one that fits best. Who is right?
    A. A only
    B. B only
    C. Both A and B
    D. Neither A nor B (E.2)

39. Undercar Specialist A says it is against federal law to tamper or disable emission controls. Undercar Specialist B says tampering, removing, or disabling emission controls is illegal in all states. Who is right?
    A. A only
    B. B only
    C. Both A and B
    D. Neither A nor B (A.1.4)

40. Undercar Specialist A says if the catalytic converter is not reducing emissions properly, the voltage frequency increases on the downstream heated oxygen sensor. Undercar Specialist B says if a fault occurs in the catalyst monitor system on three drive cycles, the MIL will be illuminated. Who is right?
    A. A only
    B. B only
    C. Both A and B
    D. Neither A nor B (B.5)

41. Technician A says that bend failures can be caused by incorrect pipe thickness. Technician B says that bend failures can be caused by improper set ups. Who is right?
    A. A only
    B. B only
    C. Both A and B
    D. Neither A nor B (C.1.4)

42. Which of the following statements about federal requirements regarding emission controls is NOT true?
    A. During the first 8 years/80,000 miles, the performance warranty covers any repair or adjustment that is necessary to make the vehicle pass an approved, locally required emission test as long as the vehicle has not exceeded the warranty time or mileage limitations and has been properly maintained according to the manufacturer's specifications.
    B. The design and defect warranty covers repair of emission-related parts that become defective during the warranty period.
    C. The design and defect warranty for model year 1995 and newer light-duty cars and trucks covers emission-control and emission-related parts for the first 2 years or 24,000 miles of vehicle use and specified major emission control components for the first 8 years or 80,000 miles of vehicle use.
    D. According to federal law, an emission-control or emission-related part, or a specified major emission control component, that fails because of a defect in materials or workmanship must be repaired or replaced by the vehicle manufacturer free of charge as long as the vehicle has not exceeded the warranty time or mileage limitations for the failed part. (E.1)

43. Undercar Specialist A says a heated air inlet controls the temperature of the air on its way to the carburetor or fuel injection body. By warming the air, it reduces HC and CO emissions by improved fuel vaporization and faster warm-up. Undercar Specialist B says some vehicles are equipped with an exhaust manifold heat control valve that routes exhaust gases to warm the intake manifold when the engine is cold. The result is reduced HC and CO emissions. Who is right?
    A. A only
    B. B only
    C. Both A and B
    D. Neither A nor B (A.1.3)

44. Undercar Specialist A says replacement catalytic converters must have documents or labels that show they meet EPA requirements and are warranted to meet federal durability and performance standards. Undercar Specialist B says if you bypass a damaged or ineffective catalytic converter instead of replacing it or repairing it, you are breaking the law and could be subject to a stiff fine. Who is right?
    A. A only
    B. B only
    C. Both A and B
    D. Neither A nor B (B.6)

45. Which of the following is NOT a true statement about the flame of an oxyacetylene torch?
    A. A carburizing flame is blue with an orange and red end.
    B. A carburizing flame results from an excess of oxygen in the mixture and tends to burn the metal being welded.
    C. A neutral flame has a quiet blue-white inner cone.
    D. A neutral flame is the result of the correct amount of acetylene and oxygen. (C.2.2)

46. Undercar Specialist A says the accuracy of the oxygen sensor reading can be affected by air leaks in the intake or exhaust manifold. Undercar Specialist B says a misfiring spark plug that allows unburned oxygen to pass into the exhaust causes the sensor to give a false lean reading. Who is right?
    A. A only
    B. B only
    C. Both A and B
    D. Neither A nor B (A.1.5)

47. The converter is one of the components covered by federal emission regulations. Which one of these statements about converter replacement is NOT true?
    A. A converter cannot be replaced by anyone other than a car dealer.
    B. Up to model year 1995, converters were covered by a 5-year/50,000-mile federal emissions warranty (7 years or 70,000 miles in California).
    C. In 1995, the warranty jumped to 8 years and 80,000 miles.
    D. If you replace a converter, you must also obtain the customer's authorization for repairs in writing; keep the paperwork for six months and the old converter for 15 days. (E.3)

48. Undercar Specialist A says that on OBD-II vehicles, the MIL will illuminate if the same fault is detected during consecutive drive cycles. Undercar Specialist B says that the DTC will be erased from memory after 40 warm-up cycles after the MIL goes out. Who is right?
    A. A only
    B. Specialist B only
    C. Both A and B
    D. Neither A nor B (A.1.3)

49. Which of the following statements about using oxygen and acetylene to cut through a pipe is NOT true?
    A. A high-pressure jet of oxygen from the cutting torch is directed at the metal that causes the metal to burn and blow away very rapidly.
    B. The proper size-cutting tip is very important.
    C. The preheat flame must provide just the right amount of heat and the oxygen jet orifice must deliver the correct amount of oxygen at the correct pressure and speed to provide a good cut.
    D. Cutting oxygen exits from a separate nozzle and jet orifice when the welder depresses the cutting oxygen lever. (C.2.4)

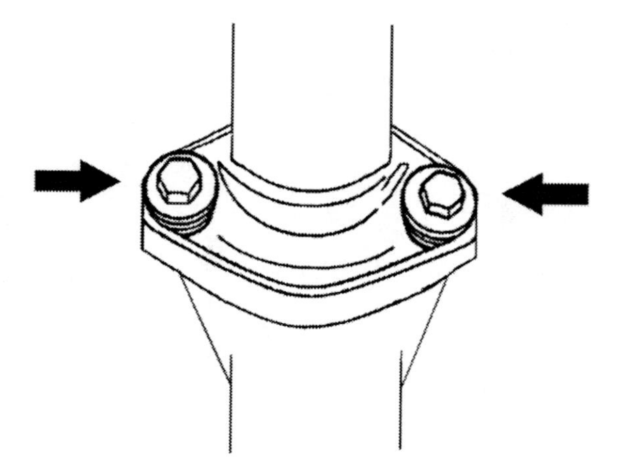

50. Undercar Specialist A says the bolts in the figure should be tightened until the spring under the bolt head is totally compressed. Undercar Specialist B says the bolts are part of a hanger assembly. Who is right?
    A. A only
    B. B only
    C. Both A and B
    D. Neither A nor B (D.3)

51. Undercar Specialist A says if the turbo sound cycles or changes in intensity, a likely cause is loose material in the compressor inlet ducts. Undercar Specialist B says a likely cause for turbocharger failure is the presence of foreign materials in the turbo unit. Who is right?
    A. A only
    B. B only
    C. Both A and B
    D. Neither A nor B (A.1.3)

52. While discussing federally regulated emission control devices, Undercar Specialist A says the common pre-combustion emission control systems, that by law must be repaired or replaced to function as they were when the vehicle was new, are the positive crankcase ventilation (PCV) and the exhaust gas recirculating (EGR) systems. Undercar Specialist B says the common post-combustion emission control systems, that by law must be repaired or replaced to function as they were when the vehicle was new, are the air management (AIR) systems, catalytic converter, and the turbocharger systems. Who is right?
    A. A only
    B. B only
    C. Both A and B
    D. Neither A nor B (E.5)

6 Additional Test Questions for Practice

Additional Test Questions

Please note the letter and number in parentheses following each question. They match the task in Section 4 that discusses the relevant subject matter. You may want to refer to the overview using the cross-referencing key to help with questions posing problems for you.

1. An exhaust system inspection should include all of the following **EXCEPT:**
   A. listening for hissing or rumbling that would result from a leak in the system.
   B. removing the catalytic converter and seeing if compressed air will pass through it.
   C. checking the system for separated connections.
   D. paying attention to the color of the catalytic converter. (A.1.1)

2. Undercar Specialist A says the exhaust manifold gasket seals the joint between the exhaust manifold and the exhaust pipe. Undercar Specialist B says that a resonator helps to reduce exhaust noise. Who is right?
   A. A only
   B. Specialist B only
   C. Both A and B
   D. Neither A nor B (A.2.1)

3. While discussing the proper way to test a catalytic converter, Undercar Specialist A says a pressure gauge can be inserted into the oxygen sensor bore to measure exhaust backpressure. Undercar Specialist B says restrictions in the converter can be checked with a vacuum gauge while the engine is being held at a fast idle. Who is right?
   A. A only
   B. B only
   C. Both A and B
   D. Neither A nor B (B.1)

4. Undercar Specialist A says the air pump in an AIR system sends air to the exhaust manifold during engine warm-up. Undercar Specialist B says the air pump sends air to the converter when the engine reaches normal operating temperatures. Who is right?
   A. A only
   B. B only
   C. Both A and B
   D. Neither A nor B (A.1.3)

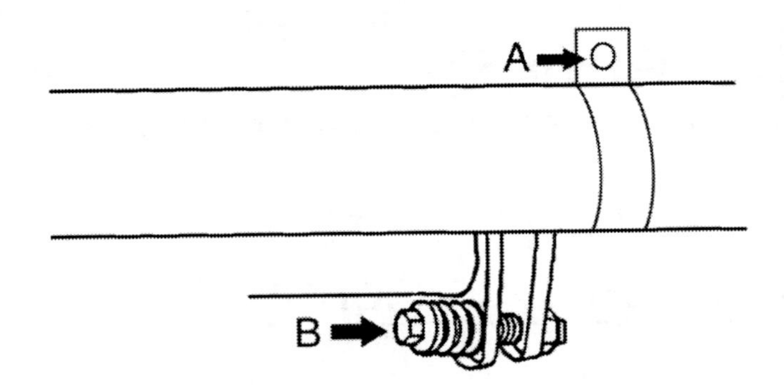

5. All of the following statements about the figure are true **EXCEPT:**
   A. part A is a U-clamp.
   B. part A is part of a hanger.
   C. part B is a spring-loaded flange bolt.
   D. part B is designed to allow for heat expansion. (A.2.4)

6. Undercar Specialist A says a turbocharger has its own, self-contained lubrication system. Undercar Specialist B says a turbocharger should not be operated at low engine oil pressure. Who is right?
   A. A only
   B. Specialist B only
   C. Both A and B
   D. Neither A nor B (A.2.5)

7. Undercar Specialist A says that AIRB valve directs secondary air either to the exhaust manifold or to the catalytic converter. Undercar Specialist B says that secondary air may be vented during deceleration. Who is right?
   A. A only
   B. B only
   C. Both A and B
   D. Neither A nor B (B.2)

8. Undercar Specialist A says MIG is not typically used with stainless steel and must not be used on high-strength steel or aluminum. Undercar Specialist B says oxyacetylene welding can be used on all metals. Who is right?
   A. A only
   B. B only
   C. Both A and B
   D. Neither A nor B (C.2.1)

9. .Undercar Specialist A says with a vacuum gauge connected to an intake manifold source and the engine at a moderate speed, the vacuum reading will be high and stay there if the exhaust is restricted. Undercar Specialist B says while touching the probe of a pyrometer on the exhaust pipe just ahead of and just behind the converter, if the inlet and outlet temperatures are the same, the converter is plugged. Who is right?
   A. A only
   B. B only
   C. Both A and B
   D. Neither A nor B (A.1.2)

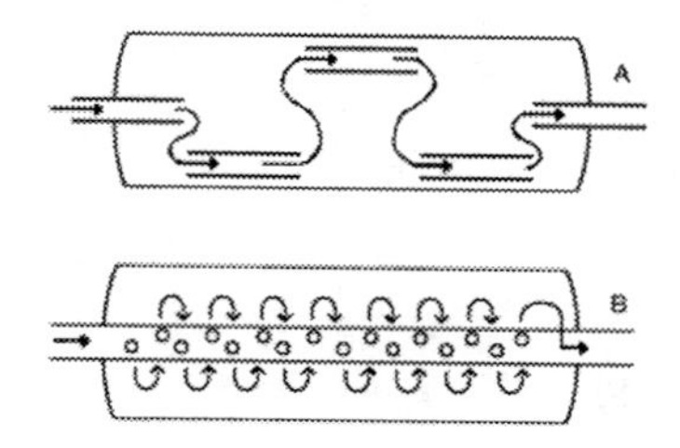

10. Undercar Specialist A says the design marked "A" in the figure is a reverse flow muffler. Undercar Specialist B says the design marked "B" in the figure is a straight-through muffler. Who is right?
    A. A only
    B. B only
    C. Both A and B
    D. Neither A nor B (D.2)

11. While checking the oxygen sensors on a late-model vehicle, Undercar Specialist A says OBD-II vehicles use a minimum of two oxygen sensors: one for fuel control and the other to give an indication of the efficiency of the converter. Undercar Specialist B says a lab scope can be used to check the oxygen sensors, as well as the converter. Who is right?
    A. A only
    B. B only
    C. Both A and B
    D. Neither A nor B (B.5)

12. While welding, Undercar Specialist A wears a welding helmet equipped with a #4 tinted filter lens while MIG welding. Undercar Specialist B wears a helmet with a #6 tinted filter lens while welding with oxyacetylene. Who is right?
    A. A only
    B. B only
    C. Both A and B
    D. Neither A nor B (C.2.2)

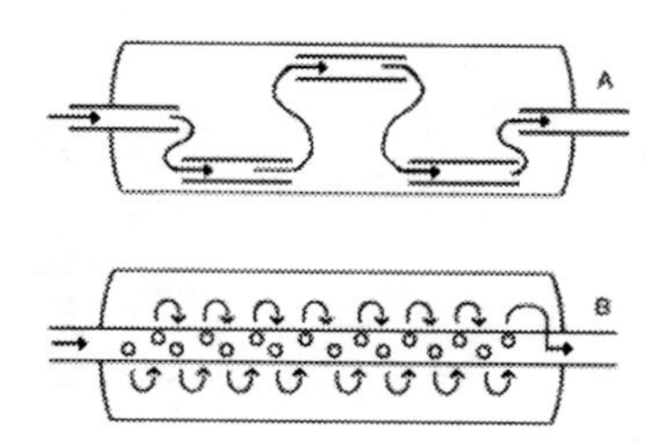

13. All of the following statements about the items shown in the figure are true **EXCEPT:**
    A. B is a straight-through muffler.
    B. in B, a straight path for the gases extends from the front to the rear of the unit through a centrally located pipe with holes.
    C. A is a muffler type that reverses the flow of the exhaust gases.
    D. A is a staggered flow muffler that relies on the random flow of exhaust to silence it. (D.2)

14. A restricted exhaust system can cause:
    A. stalling.
    B. loss of power.
    C. backfiring.
    D. All of the above (A.1.2)

15. Which of the following would NOT be considered tampering or disabling an emission control device?
    A. The presence of a straight pipe where the catalytic converter should be
    B. The enlargement of the fuel filler neck so that it can accept leaded fuels
    C. The presence of larger diameter exhaust pipes
    D. The plugging of the vacuum lines to the EGR valve (A.1.4)

16. All of the following are typical emission control systems **EXCEPT:**
    A. EGR
    B. PCV
    C. EPA
    D. EFE (A.1.3)

17. While discussing catalytic converter diagnosis, Undercar Specialist A says a delta temperature test should be conducted. Undercar Specialist B says a good converter is evident by low amounts of CO_2 in the exhaust. Who is right?
    A. A only
    B. B only
    C. Both A and B
    D. Neither A nor B (B.1)

18. Undercar Specialist A says the OBD-II heated oxygen sensor monitor can detect engine misfires. Undercar Specialist B says that it can monitor catalytic converter efficiency. Who is right?
    A. A only
    B. B only
    C. Both A and B
    D. Neither A nor B (B.5)

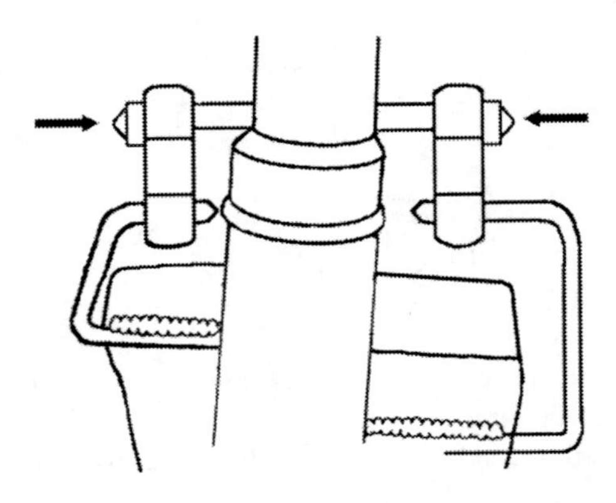

19. To install the items shown in the figure,:
    A. a hammer must be used to drive them into place.
    B. a soapy solution should be put over the metal parts so the rubber can slip over.
    C. a special tool is needed to press these insulators in place.
    D. the insulators must be heated in order to fit them over the metal prongs. (D.4)

20. Which emission control system introduces exhaust gases into the intake air to reduce the formation of NOx in the combustion chamber?
    A. Evaporative emission controls
    B. Exhaust gas recirculation
    C. Air injection
    D. Early fuel evaporation (A.1.3)

21. All of the following statements are true **EXCEPT:**
    A. the accuracy of the oxygen sensor reading can be affected by air leaks in the exhaust manifold.
    B. a misfiring spark plug that allows unburned oxygen to pass into the exhaust also causes the sensor to give a false rich reading.
    C. if the heated oxygen sensor wiring, connector, or terminal is damaged, the entire oxygen sensor assembly should be replaced.
    D. in order for this sensor to work properly, it must have a clean air reference, which it receives from the air that is present around the sensor's signal and heater wires. (A.1.5)

22. What is the first step in turbocharger inspection?
    A. Check the air cleaner for a dirty element.
    B. Open the turbine housing at both ends.
    C. Start the engine and listen to the system.
    D. Remove the ducting from the air cleaner to turbo and examine the area. (A.2.5)

23. Undercar Specialist A says a restricted exhaust passage in an EGR valve may cause the engine to run rough or stall at idle. Undercar Specialist B says the EGR passages are not restricted if the engine stalls at idle when the EGR is fully opened. Who is right?
    A. A only
    B. B only
    C. Both A and B
    D. Neither A nor B (B.3)

24. All of the following statements about pulse-type AIR systems are true **EXCEPT:**
    A. this system uses the natural exhaust pressure pulses to pull air from the air cleaner into the exhaust manifolds and/or the catalytic converter.
    B. at lower engine speeds, each negative pressure pulse in the exhaust manifold moves air from the air cleaner into an exhaust port.
    C. high-pressure pulses in the exhaust manifold close the one-way check valves and prevent exhaust from entering the system.
    D. this system is more effective in reducing HC emissions at higher engine speeds. (A.1.3)

25. Which of the following statements about the heated oxygen sensor monitor function of an OBD-II system is NOT true?
    A. The system monitors lean to rich and rich to lean time responses.
    B. The system can detect a lazy oxygen sensor that cannot switch fast enough to keep proper control of the air-fuel mixture in the system.
    C. The heated oxygen sensor monitor checks the voltage signal frequency of the upstream oxygen sensor to determine the condition of all oxygen sensors in the system.
    D. At certain times, the heated oxygen sensor monitor varies the fuel delivery and checks for oxygen sensor response. (B.5)

26. Before beginning to set or light an oxyfuel torch, Undercar Specialist A turns the regulator adjusting screws all the way out before opening the cylinder valves. Undercar Specialist B purges the lines after setting the cylinder regulators. Who is right?
    A. A only
    B. B only
    C. Both A and B
    D. Neither A nor B (C.2.1)

27. Undercar Specialist A says an oxygen sensor can be a voltage-producing sensor. Undercar Specialist B says an oxygen sensor is a thermistor sensor. Who is right?
    A. A only
    B. B only
    C. Both A and B
    D. Neither A nor B (A.1.5)

28. Undercar Specialist A says if vacuum is applied to a positive back pressure EGR valve when the engine is off, a good valve will open. Undercar Specialist B says if vacuum is applied to a negative back pressure EGR valve when the engine is off, a good valve will open. Who is right?
    A. A only
    B. Specialist B only
    C. Both A and B
    D. Neither A nor B (A.1.3)

29. Undercar Specialist A says the gasket used to seal the part shown in the figure is necessary only when the machined surface is warped. Undercar Specialist B says if the part is warped beyond manufacturer's specifications or is cracked, it must be replaced. Who is right?
    A. A only
    B. Specialist B only
    C. Both A and B
    D. Neither A nor B (A.2.3)

30. Which of the following statements is NOT true about MIG welding?
    A. Prior to filling in a weld, tack the parts to be welded into their proper position.
    B. The arc length is determined by the voltage; therefore the welder must watch and control the distance from the nozzle to the work.
    C. By controlling the nozzle-to-work distance, the welder will control the voltage at the weld.
    D. The welding speed and torch angle determines the bead width and appearance. (C.2.1)

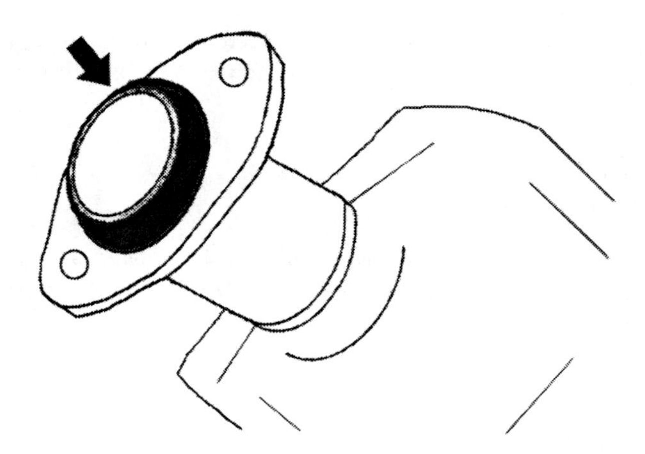

31. Undercar Specialist A says the part shown in the figure is replaced whenever the converter is replaced. Undercar Specialist B says the part should be replaced whenever the flange at the converter is loosened or disconnected. Who is right?
    A. A only
    B. B only
    C. Both A and B
    D. Neither A nor B
    (D.3)

32. While discussing OBD-II systems, Undercar Specialist A says the PCM illuminates the MIL if a defect causes emissions levels to exceed 2.5 times the emission standards for that model year vehicle. Undercar Specialist B says if a misfire condition threatens engine or catalyst damage, the PCM flashes the MIL. Who is right?
    A. A only
    B. Specialist B only
    C. Both A and B
    D. Neither A nor B
    (B.1)

33. All of the following statements are true **EXCEPT:**
    A. to provide the right amount of gas to a torch tip with a larger orifice, lower pressures are set.
    B. regardless of the torch tip or the pressures used, if the flame is neutral, it will have the same temperature.
    C. if the torch tip orifice is too small, not enough heat will be available to bring the metal to its melting and flowing temperature.
    D. if the torch tip is too large, poor welds will result because the weld will need to be made too quickly, the welding rod will melt too quickly, and the weld pool will be hard to control.
    (C.2.2)

34. Undercar Specialist A says backfiring through the exhaust on a vehicle equipped with an AIR system can be caused by a bad diverter valve in the system. Undercar Specialist B says the one-way check valve can be checked with a hand-held vacuum pump. Who is right?
    A. A only
    B. B only
    C. Both A and B
    D. Neither A nor B
    (A.1.3)

35. While discussing EGR valve diagnosis, Undercar Specialist A says if the EGR valve does not open, the engine may hesitate on acceleration. Undercar Specialist B says if the EGR valve does not open, the engine may detonate on hard acceleration. Who is right?
    A. A only
    B. B only
    C. Both A and B
    D. Neither A nor B (B.3)

36. While discussing oxygen sensor diagnosis, Undercar Specialist A says the voltage signal on a satisfactory O_2S should cycle between 0.5V and 1.0V. Undercar Specialist B says it should cycle between 0.5V and 0V. Who is right?
    A. A only
    B. B only
    C. Both A and B
    D. Neither A nor B (B.5)

37. Which of the following is NOT a true statement about pipe bending?
    A. The straight pipe is laid out against the original and bends are made one at a time until the pipe matches the original.
    B. Some pipe-bending machines use program cards to bend the pipe at preset angles for the vehicle.
    C. Once the correct bends are made to the pipe, the pipe should be twisted so that the bends are at the right angle.
    D. When making a bend in the pipe, make sure the depth of the bend is exactly what is required. (C.1.2)

38. Undercar Specialist A says oxygen sensors produce a voltage based on the amount of oxygen in the exhaust. Undercar Specialist B says rich mixtures release large amounts of oxygen in the exhaust; therefore the oxygen sensor voltage is high. Who is right?
    A. A only
    B. B only
    C. Both A and B
    D. Neither A nor B (A.1.3)

39. Which of the following statements about MIG welding is NOT true?
    A. Direct current power sources used for MIG welding typically have the wire (electrode) as the negative and the workpiece as positive; weld penetration is greater using this connection.
    B. The use of CO_2 gas allows for the best weld penetration.
    C. CO_2 as a shielding gas gives a harsher, unstable arc, which leads to increased spatter.
    D. Voltage adjustment and wire feed speed must be set according to the diameter of the wire being used and metal thickness. (C.2.2)

40. A mini-converter/preheater is used:
    A. on small vehicles where a normal converter will not fit properly.
    B. on engines that use leaded fuels.
    C. in conjunction with EGR systems to supply clean exhaust for the cylinders.
    D. to reduce emissions during engine warm-up. (A.2.2)

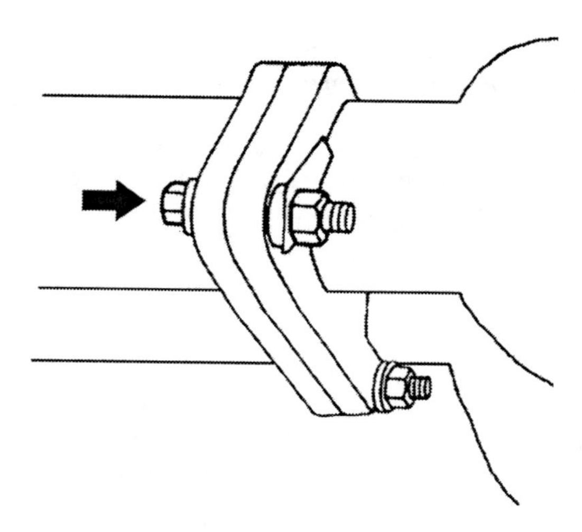

41. While attempting to loosen the bolts shown in the figure, Undercar Specialist A uses an air chisel to knock the head off the bolt. Undercar Specialist B puts a wrench on each end of the bolt and tries to tighten it before loosening it. Who is right?
    A. A only
    B. B only
    C. Both A and B
    D. Neither A nor B
    (D.4)

42. Undercar Specialist A says the emissions warranties apply to used vehicles, as well as new ones, as long as the vehicle has not exceeded the warranty time or mileage limitations. Undercar Specialist B says if a vehicle has had all the proper maintenance performed and it still fails an emissions test, the catalytic converter should be replaced. Who is right?
    A. A only
    B. B only
    C. Both A and B
    D. Neither A nor B
    (E.1)

43. Which of the following statements about an MIL on a late-model vehicle is NOT true?
    A. When a misfire that threatens the engine or can cause catalyst damage occurs, the misfire monitor flashes the MIL on the first occurrence of the misfire.
    B. A fault detected by the catalyst monitor flashes the MIL on the first occurrence of the mishap.
    C. For the catalyst efficiency, HO_2, EGR, and comprehensive component monitors, the MIL is turned off if the same fault does not reappear for three consecutive drive cycles.
    D. When the fault is no longer present and the MIL is turned off, the DTC is erased after 40 engine warm-up cycles.
    (A.1.3)

44. While checking engine/transmission mounts to identify the cause of exhaust system breakage, Undercar Specialist A pulls up on the engine and says if the mount's rubber separates from the metal plate, the mount should be replaced. Undercar Specialist B pulls and pushes down on the engine and transmission and says if there is movement between the metal plate of the mount and its attaching point on the frame, tighten the attaching bolts to the appropriate torque and recheck. Who is right?
    A. A only
    B. B only
    C. Both A and B
    D. Neither A nor B (A.1.6)

45. When discussing the diagnosis of a positive backpressure EGR valve, Undercar Specialist A says with the engine running at idle speed, if a hand pump is used to supply vacuum to the EGR valve, the valve should open at 12 in. Hg of vacuum. Undercar Specialist B says with the engine not running, any vacuum supplied to the EGR valve should be bled off and the valve's diaphragm should not move. Who is right?
    A. A only
    B. Specialist B only
    C. Both A and B
    D. Neither A nor B (B.3)

46. While checking the activity of an oxygen sensor on a scanner with the engine running, Undercar Specialist A says the oxygen sensor voltage should quickly cycle between zero and one volt. Undercar Specialist B says the oxygen sensor should be cool to the touch during the scan tests. Who is right?
    A. A only
    B. B only
    C. Both A and B
    D. Neither A nor B (B.5)

47. Which of the following systems is designed to reduce NOx and has little or no effect on overall engine performance?
    A. PCV
    B. EGR
    C. AIR
    D. EVAP (A.1.3)

48. Which of the following statements is NOT true about installing a turbocharger unit?
    A. When installing a new or remanufactured turbocharger, make certain that the oil inlet and drain lines are clean before connecting them.
    B. Be sure the engine oil is clean and add two extra quarts of oil to the engine; fill the oil filter with clean oil.
    C. Leave the oil drain line disconnected at the turbo and crank the engine without starting it until oil flows out of the turbo drain port.
    D. Connect the drain line, start the engine, and operate it at low idle for a few minutes before running it at higher speeds. (A.2.5)

49. Undercar Specialist A listens to sounds the turbo makes and says a high-pitched sound is normal. Undercar Specialist B says if the turbocharger's sound changes intensity, there may be dirt built up on the turbocharger's compressor wheel. Who is right?
    A. A only
    B. Specialist B only
    C. Both A and B
    D. Neither A nor B
    (A.1.3)

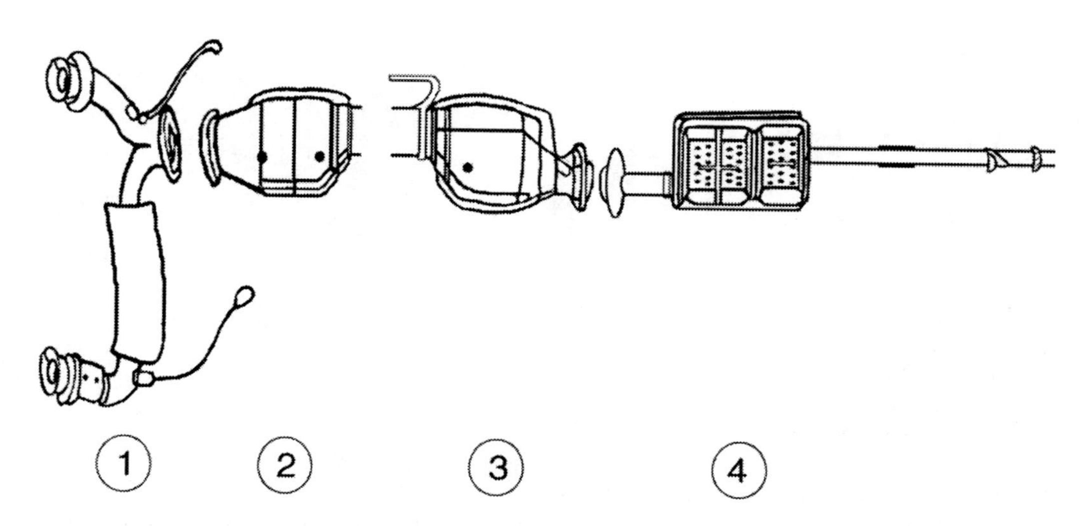

50. Part 2 in the figure is MOST Likely a:
    A. catalytic converter.
    B. muffler.
    C. resonator.
    D. mini-converter/preheater.
    (A.2.1)

51. All of the following are acceptable methods to verify the integrity of a weld **EXCEPT:**
    A. visual inspection.
    B. pressure test (plug exhaust).
    C. vacuum test.
    D. noise test.
    (C.2.1)

52. If a vehicle is within the age and mileage limits for the applicable emissions warranty, an automobile manufacturer can only deny coverage if evidence shows that the owner has failed to properly maintain and use the vehicle, causing the part or emission test failure. Which of the following would not be considered an example of misuse and lack of proper maintenance?
    A. Off-road driving or overloading
    B. Use of high-octane fuels
    C. Tampering with emission control parts or systems
    D. Improper maintenance
    (E.1)

53. Which of the following is NOT a recommended way to cut through an exhaust pipe?
    A. Manual pipe cutter
    B. Oxyacetylene torch
    C. Plasma arc
    D. Metal inert gas
    (C.2.3)

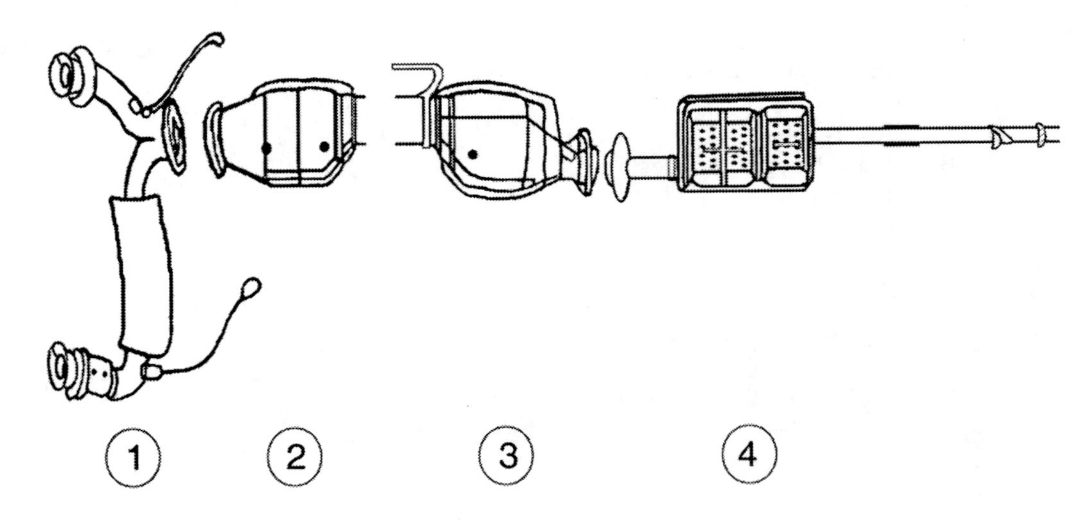

54. Part 1 in the figure is a(n):
    - A. catalytic converter.
    - B. muffler.
    - C. resonator.
    - D. oxygen sensor. (A.2.1)

55. Which of the following statements about servicing heated oxygen sensors is NOT true?
    - A. The sensor pigtail and harness wires must not be damaged in such a way that the wires inside are exposed. This could provide a path for foreign materials to enter the sensor and cause performance problems.
    - B. If the sensor or wiring harness connectors are dirty, they should only be cleaned with electrical contact cleaner.
    - C. Neither the sensor nor vehicle lead wires should be bent sharply or kinked. Hard bends, kinks, etc., could block the reference air path through the lead wire.
    - D. Never remove or defeat the oxygen sensor ground wire. Vehicles that utilize the ground wired sensor may rely on this ground as the only ground contact to the sensor. Removal of the ground wire will cause poor engine performance. (A.1.3)

Appendices

Answers to the Test Questions for the Sample Test Section 5

1. C	14. A	27. B	40. C
2. B	15. A	28. A	41. C
3. C	16. C	29. A	42. A
4. D	17. D	30. C	43. C
5. A	18. C	31. C	44. C
6. A	19. A	32. A	45. B
7. B	20. D	33. C	46. C
8. C	21. B	34. A	47. A
9. C	22. B	35. C	48. B
10. C	23. B	36. A	49. D
11. C	24. A	37. D	50. D
12. B	25. C	38. A	51. C
13. D	26. A	39. C	52. A

Explanations to the Answers for the Sample Test Section 5

Question #1
Answer A is wrong. Oxyacetylene welding is directly controlled by the welder.
Answer B is wrong. The one touch controls the flow of oxygen to the flame to increase flame temperatures.
Answer C is correct. Oxyacetylene welding provides a high-temperature flame for welding by combining oxygen and acetylene gases. The flame of this combination provides enough heat to melt most metals. Oxyacetylene welding is a manual process; the welder controls the torch movement and the welding rod. The torch is connected via hoses to separate cylinders of gas. The flow rate from the cylinders to the torch is controlled by a manifold gauge set at the tanks and by needle valves at the torch. One torch valve controls the rate of oxygen flow and the other controls the rate of acetylene flow. The two gases mix and burn at the orifice of the torch, not in the handle of the torch as stated in answer C. Oxyacetylene welding joins metals by melting and fusing. A very intense, concentrated flame is applied to the metal until a spot under the flame becomes molten and forms a pool of liquid metal. When two metals melt and the molten pools run together and fuse, the edges of the two become one. When you want to weld on the exhaust systems of cars that have been made in the last 10 years, you need to determine the types of metal from which they are made. Unlike older cars, whose exhaust systems were mostly made of mild steel tubing, the metals used today are a mixture of different types of metals. The two most common mixtures are a combination of aluminum and milled steel and stainless steel and milled steel. Both of these mixes help cut down on rust and corrosion; yet, for the technician making the repairs without some special training, they pose a problem. The gas welding of these pipes takes practice and a colder flame to keep the metal from getting brittle and breaking.
Answer D is wrong. Welding is for joining metals together.

Question #2
Answer A is wrong. Preheating the entire cut line before cutting helps the process.
Answer B is correct. Only Undercar Specialist B is correct. Undercar Specialist A is wrong. To cut, bring the tip of the inner cone of the preheat flame to the edge of the metal to be cut. The torch should be held so that the inner cone of the preheat flame is about 1⁄16 to 1⁄8 inch away from the surface of the metal. As soon as a spot on the cutting line has been heated to a bright cherry red color, squeeze the cutting oxygen lever. The jet of oxygen coming from the tip will cause the heated metal to burn away.
Answer C is wrong. Only Undercar Specialist B is correct.
Answer D is wrong. Only Undercar Specialist B is correct.

Question #3
Answer A is wrong. Undercar Specialist B is also correct.
Answer B is wrong. Undercar Specialist A is also correct.
Answer C is correct. Both Undercar Specialists are correct. After the exhaust has passed through a good catalytic converter, only harmless gases should be emitted. The use of leaded fuel and overheating are the primary causes of converter failure. The catalytic converter is one of the most important components for emissions control. It is also one of the most fragile components and the easiest to damage from misuse. Most modern cars are equipped with three-way catalytic converters. "Three-way" refers to the three regulated emissions it helps to reduce: carbon monoxide, HC, and NO_x molecules. The converter uses two different types of catalysts: a reduction catalyst and an oxidization catalyst. Both types consist of a ceramic structure coated with a metal catalyst, usually platinum, rhodium, and/or palladium. The idea is to create a structure that exposes the maximum surface area of catalyst to the exhaust stream. There are two main types of structures used in catalytic converters: honeycomb and ceramic beads. Most cars today use a honeycomb structure.
Answer D is wrong. Both Undercar Specialists are correct.

Question #4
Answer A is wrong. A stuck valve is caused by carbon build-up problems.
Answer B is wrong. Diaphragms get damaged with age and weather deterioration.
Answer C is wrong. Carbon excessive build up can be an indication of other engine problems.
Answer D is correct, because it is wrong. Detonation or spark knock can be caused by any condition that prevents proper EGR gas flow, such as a valve stuck closed, leaking valve diaphragm, restrictions in flow passages, EGR disconnected, or a problem in the vacuum source. An open EGR valve can cause a no start condition, surging, or stalling. To determine the causes of spark knock from the EGR system problem, first look for obvious problems, such as broken or loose vacuum hoses. Then, test the system for proper operation of the valve, all of the components in the system, and computers that may render the EGR system inoperative. Once you verify the other systems are working properly, check the EGR system to see if the vacuum is getting to the valve when it is supposed to.

Question #5
Answer A is correct. Only Undercar Specialist A is correct. Exhaust systems are designed for the vehicle they are in and the engine that is in the vehicle. In-line engines have a simple design that runs down one side of the engine. Called a single exhaust system, there is only one pipe to the rear of the car. Single exhausts can be found on all engine sizes. A single exhaust system has one path for exhaust flow through the system. Typically, it has only one header pipe, a main catalytic converter, a muffler, and a tailpipe. V-type engines with dual exhaust have two pipes, two mufflers, and two catalytic converters. A dual exhaust system has two separate exhaust paths to reduce back pressure. A crossover pipe joins the two manifolds into one pipe as they leave the engine on a V-type engine's single exhaust system. A crossover pipe normally connects the right and left header pipes to equalize exhaust pulses in a true dual system. This also increases engine power slightly.
Answer B is wrong. Dual exhaust starts from each manifold and is separate from the start.
Answer C is wrong. Only Undercar Specialist A is correct.
Answer D is wrong. Only Undercar Specialist A is correct.

Question #6
Answer A is correct. Only when conducting a visual inspection of an exhaust system, raise the car on a lift. Using a droplight, closely inspect the system for problems. Inspect all exhaust system joints for leaks. Look for soot and discoloration, which indicate escaping hot gases. These leaks must be corrected to enable the vehicle to pass emissions testing and, more importantly, to protect the vehicle's occupants from being exposed to toxic gases. Pay particular attention to the muffler and all pipe connections, gaskets, and pipe bends. Exhaust leaks will often show up as gray or white carbon lines coming from openings. Catalytic converters can overheat and will leave telltale signs of doing so. Look for bluish or brownish discoloration of the outer stainless steel shell. Also, look for blistered or burnt paint or undercoating above and near the converter.
Answer B is wrong. Indications of converter over heating can cause internal damage and failure.
Answer C is wrong. Only Specialist A is correct.
Answer D is wrong. Only Specialist A is correct.

Question #7
Answer A is wrong. The EGR will cause a code specific to it.
Answer B is correct. The possible causes for OBD-II AIR monitor failure are faulty secondary AIR solenoid and/or relay; damaged, loose, or disconnected wiring in the secondary air solenoid and/or relay circuit; defective aspirator valve; disconnected or damaged AIR hoses and/or tubes; a defective electric or mechanical air pump; air pump drive belt missing; and a faulty AIR check valve.
Answer C is wrong. The HO_2S will create code to fuel control.
Answer D is wrong. A faulty PCV valve may cause codes to fuel control.

Question #8
Answer A is wrong. Undercar Specialist A is also correct.
Answer B is wrong. Undercar Specialist B is also correct.
Answer C is correct. Both Undercar Specialists are correct. If the vacuum routing for the AIR system is not correct and air is diverted upstream when the vehicle enters closed loop operation, a false lean mixture would be sensed by the O_2 sensor, causing a rich mixture command by PCM. Undercar Specialist A is correct. Rich mixtures can cause the catalytic converter to run hot and overheat. Undercar Specialist B is also correct.
Answer D is wrong. Both Undercar Specialists are correct.

Question #9
Answer A is wrong. Undercar Specialist B is also correct.
Answer B is wrong. Undercar Specialist A is also correct.
Answer C is correct. Both Undercar Specialists are correct. Undercar Specialist A is correct. On some systems, the TP sensor informs the PCM as to when the engine is decelerating or accelerating. If the TP sensor were bad, EGR valve control would be affected. Likewise, Undercar Specialist B is also correct. The ECT tells the PCM when the engine is warmed up and ready for EGR action. If the ECT is bad, the PCM may activate EGR action when the engine is cold. This would affect drivability. The EGR valve on most vehicles works only when the vehicle is in a cruise condition or steady throttle position above idle, and only when the engine is warm and at operating temperature.
Answer D is wrong. Both Undercar Specialists are correct.

Question #10
Answer A is wrong. Undercar Specialist B is also correct.
Answer B is wrong. Undercar Specialist A is also correct.
Answer C is correct. Both Undercar Specialists are correct. If a converter has a bluish or brownish color, this is an indication that the converter has overheated. Therefore Undercar Specialist A is right. So is Undercar Specialist B. An ignition misfire will cause overheating and damage to the catalytic converter due to the high concentration of hydrocarbons (unburned fuel) in the exhaust. The hydrocarbons burn in the head pipe and converter causing temperatures to exceed the 600 degrees at which the converter is designed to operate. When you see the catalytic converter change to a bluish color, you need to look into the reason the engine is running richer than it should. A rich mixture causes the converter to get hotter from working harder to burn off the extra fuel; this causes the change in color. First, scan for computer codes to see if there is a problem indicated. If you only get rich codes, you should perform the pinpoint test for that code. Otherwise, check for a misfire and for excessive fuel pressure. Then, do a backpressure check on the catalytic converter.
Answer D is wrong. Both Undercar Specialists are correct.

Question #11
Answer A is wrong. Undercar Specialist B is also correct.
Answer B is wrong. Undercar Specialist A is also correct.
Answer C is correct. Both Undercar Specialists are correct. It is advisable that before beginning work on an exhaust system, make sure it is cool to the touch. Some technicians disconnect the battery's negative cable before starting to work to avoid short-circuiting the electrical system. Soak all rusted bolts, nuts, etc., with a good penetrating oil. Finally, check the system for critical clearance points so they can be maintained when new components are installed. Before attempting any repairs on an exhaust system, look at the positions of the pipes and components. If you don't observe this beforehand, you may have a clearance problem with the system. It is a good practice to use a high temperature, anti-seize compound on all the bolts and nuts used to prevent them from binding and possibly breaking during installation.
Answer D is wrong. Both Undercar Specialists are correct.

Question #12
Answer A is wrong. This would indicate a blockage in the converter.
Answer B is correct. To check the activity of a catalytic converter, conduct the delta temperature test. To conduct this test, use a hand-held digital pyrometer. By touching the pyrometer probe to the exhaust pipe just ahead of and just behind the converter, there should be an increase of at least 100° F or 8 percent above the inlet temperature reading as the exhaust gases pass through the converter. If the outlet temperature is the same or lower, nothing is happening inside the converter. This is a general test, and results may vary from car to car; so, use it as a baseline from which to work. Before you condemn the catalytic converter, check all of the other systems first. Do this to make sure the fuel and ignition systems are working at peak performance so as not to damage the new catalytic converter when installing. The rest of the exhaust should be inspected as well, so that the exhaust is working at peak efficiency.
Answer C is wrong. This would indicate a broken substrate in the converter.
Answer D is wrong. This would indicate a blockage in the converter.

Question #13
Answer A is wrong. Clamps that are used to seal a joint are positive clamps such as a U-clamp. Therefore Undercar Specialist A is wrong. Part A is a hanger.
Answer B is wrong. Undercar Specialist B is also wrong. Part B is part of union assembly between two flanges. The spring is placed on one end to allow expansion due to heat while keeping the flange connection tight.
Answer C is wrong. Neither Undercar Specialist is correct.
Answer D is correct. Neither Undercar Specialist is correct.

Question #14
Answer A is correct. Only Undercar Specialist A is correct. When the engine is cold, vacuum is applied to the heat riser and the vacuum actuator overcomes spring tension to close the valve to assist in preheating the intake manifold. When the vacuum is removed, spring tension opens the valve to allow unrestricted exhaust flow.
Answer B is wrong. By design, the vacuum actuator defaults to an open position to allow unrestricted exhaust flow. Heat risers were used on earlier vehicles before fuel injection systems, and they helped to atomize fuel better in colder weather. The system was designed to go to an open position if the vacuum system that controlled the valve failed. For later systems with fuel injection, coolant flows through the throttle body to prevent the throttle plates from getting iced up in colder weather and getting stuck, either in the closed or open position.
Answer C is wrong. Only Undercar Specialist A is correct.
Answer D is wrong. Only Undercar Specialist A is correct.

Question #15
Answer A is correct. Only Undercar Specialist A is correct. When the turbo chargers boost, pressures exceed manufacturer's design specifications, and the wastegate is opened to relieve excess boost pressure.
Answer B is wrong. Low octane fuels burn at a faster rate than high octane fuels. The resulting high combustion chamber pressures created by increased boost, will cause an even faster burn rate of low octane fuels. This increases the possibility of detonation that could cause damage to the engine. When working with turbochargers, the boost pressures are limited to the components used in the engine; excessive pressures will cause leaks and component failure. The boost pressures are regulated by the wastegate. If you over boost an engine beyond its recommended pressures, it may cause failure of engine gaskets or components. Using a lower octane fuel may increase the chance of pre-ignition.
Answer C is wrong. Only Undercar Specialist A is correct.
Answer D is wrong. Only Undercar Specialist A is correct.

Question #16
Answer A is wrong. Undercar Specialist B is also correct.
Answer B is wrong. Undercar Specialist A is also correct.
Answer C is correct. Both Undercar Specialists are correct. Using MIG welding, the welder must select the size of the electrode then set the unit to the desired voltage. The welder also must adjust the flow of the shielding gas and the rate of electrode feed. When a Mig welder is used, it's common for the user to forget to disconnect the vehicles battery. If you forget to do this simple step, there is a possibility that the welder will send high amperes through the car's wiring, melt wires, and burn up computer components in the process.
Answer D is wrong. Both Undercar Specialists are correct.

Question #17
Answer A is wrong. A is wrong because very few shops will weld in a new converter. It is more cost effective to change the whole pipe. The presence of an upstream O_2S in the graphic indicated a full blown converter not a crossover pipe with a mini-converter.
Answer B is wrong. Undercar Specialist B is right. Undercar Specialist A is wrong; the converter is part of the assembly and is replaced as a unit.
Answer C is wrong. Neither Undercar Specialist is correct.
Answer D is correct. Neither Undercar Specialist is correct. A is wrong because very few shops will weld in a new converter. It is more cost effective to change the whole pipe. The presence of an upstream oxygen sensor in the graphic indicated a full blown converter not a crossover pipe with a mini-converter. The figure shows a crossover pipe with a mini-converter, so Undercar Specialist B is right. Undercar Specialist A is wrong; the converter is part of the assembly and is replaced as a unit. When the mini-converter in a y-pipe or crossover pipe goes bad, it's general practice to replace the complete assembly, due to the cost of installing and welding in a new one. The problem most shops have doing this is getting the header pipes to fit correctly again. At the time of manufacture, a specific converter in the pipe assembly is made; consequently, you might not be able to get the exact part. This would be in violation of the EPA Clean Air Act laws. Install a direct factory or aftermarket certified converter and pipe assembly to make sure that the repair is done correctly and to ensure that the customer will be unlikely to have another problem for the life of the vehicle.

Question #18
Answer A is wrong. Undercar Specialist B is also correct.
Answer B is wrong. Undercar Specialist A is also correct.
Answer C is correct. Both Undercar Specialists are correct. If the voltage is continually high, the air-fuel ratio may be rich. When the O_2S voltage is continually low, the air/fuel ratio may be lean, the sensor may be defective, or the wire between the sensor and the computer may have a high-resistance problem. If the O_2S voltage signal remains in a mid-range position, the computer may be in open loop or the sensor may be defective. When starting to diagnose a problem with the O_2 sensors, look at other items that may cause an O_2 to give a false reading. First, always check the fuel pressures and repair, as needed. Check the air filter and throttle body for restrictions and repair, as needed. Look at the data stream from the computer; look at the long fuel trim and short fuel trims to see if they are in the negative or positive from zero. These are the indicators of what the fuel injection and computer have been feeding the engine.
Answer D is wrong. Both Undercar Specialists are correct.

Question #19
Answer A is correct. Part 2 is a catalytic converter; therefore answer A is right.
Answer B is wrong. Part 2 is not a muffler.
Answer C is wrong. Part 2 is not a resonator.
Answer D is wrong. Part 2 is not a mini-converter.

Question #20
Answer A is wrong. If you attempt to sandpaper the blades could cause harm to the turbo.
Answer B is wrong. If you scrape the turbo blades will cause damage and unbalance the blades.
Answer C is wrong. Doing either of these repair attempts will damage the turbo.
Answer D is correct. Neither Undercar specialists is correct. You should never attempt to scrape or clean off any dirt or debris from the compressor of a turbocharger with anything but a rag. Doing anything else could throw the wheel off balance.

Question #21
Answer A is wrong. Manifolds warp often due to the different metals of the cylinder head and manifold.
Answer B is correct. The figure shows an exhaust manifold. The manifold itself rarely causes any problems. On occasion, an exhaust manifold will warp because of excess heat. A straightedge and feeler gauge can be used to check the machined surface of the manifold. Another problem that can also result from the high temperatures generated by the engine is a cracked manifold. This usually occurs after the car passes through a large puddle and cold water splashes on the manifold's hot surface. If the manifold is warped beyond manufacturer's specifications or is cracked, it must be replaced.
Answer C is wrong. If a manifold gets overheated it may crack.
Answer D is wrong. Due to a quick change in temperatures, manifolds may crack.

Question #22
Answer A is wrong. This is a true statement. If the sensor reading is out of this range it is defective.
Answer B is correct. To check an oxygen sensor with a voltmeter, connect the voltmeter between the O_2S wire and ground. Backprobe the connector near the O_2S to connect the voltmeter to the sensor signal wire. If possible, avoid probing through the insulation to connect a meter to the wire. With the engine idling, the sensor voltage should be cycling from low voltage to high voltage. The signal from most oxygen sensors varies between 0 and 1 volt. If the voltage is continually high, the air/fuel ratio may be rich or the sensor may be contaminated by RTV sealant, antifreeze, or lead from leaded gasoline. When the O_2S voltage is continually low, the air/fuel ratio may be lean, the sensor may be defective, or the wire between the sensor and the computer may have a high-resistance problem. If the O_2S voltage signal remains in a mid-range position, the computer may be in open loop or the sensor may be defective. The oxygen sensor that monitors catalytic converter operation would stay around the .5 volt level on a normal operating sensor that is behind the catalytic converter. The oxygen sensor on most vehicles will vary between .2 volts and .8 volts. So, look for the sensor to sweep back and forth while the engine is running at normal operating temperature.
Answer C is wrong. If the wrong type RTV is used this could foul the sensors.
Answer D is wrong. If the heater circuit is not working the sensor will take too long to heat up.

Question #23
Answer A is wrong. This would only be used as a last resort if no other information on the angles are available.
Answer B is correct. Only Undercar Specialist B is correct. When bending pipe for an exhaust system, the pipe can be bent to match the old or original pattern. To do this, the straight pipe is laid out against the original and bends are made one at a time until the pipe matches the original. This is a time-consuming skill, as each bend is compared and tweaked until it matches the original. Some pipe-bending machines are somewhat intelligent. These use program cards to bend the pipe at preset angles for the vehicle. Again, the pipe is bent one bend at a time until it matches the original configuration. It is important to match the location and twist of the bends. Having the correct bends is not enough; the bends must be at the right place and at the right angle.
Answer C is wrong. Only Undercar Specialist B is correct.
Answer D is wrong. Only Undercar Specialist B is correct.

Question #24
Answer A is correct. Only Undercar Specialist A is correct. The parts shown in the figure are rubber insulators. Their sole purpose is to keep the exhaust system aligned to prevent rattles and to prevent exhaust vibrations from reaching the passenger compartment. Alignment of the exhaust system can be critical, when it comes to trying to stop leaks and also to stop noises.
Answer B is wrong. If these rubber supports break could lead to exhaust damage from contact.
Answer C is wrong. Only Undercar Specialist B is correct.
Answer D is wrong. Only Undercar Specialist B is correct.

Question #25
Answer A is wrong. Undercar Specialist B is also correct.
Answer B is wrong. Undercar Specialist A is also correct.
Answer C is correct. Both Undercar Specialists are correct. The exhaust gas recirculation (EGR) system introduces exhaust gases into the intake air to reduce the temperatures reached during combustion. This reduces the chances of forming NO_x during combustion. The air-injection system (AIR) reduces HC emissions by introducing fresh air into the exhaust stream to cause minor combustion of the HC in the engine's exhaust.
Answer D is wrong. Both Undercar Specialists are correct.

Question #26
Answer A is correct because it is wrong. Heat shields are used to protect other parts from the heat of the exhaust system and the catalytic converter. They are usually made of pressed or perforated sheet metal. Heat shields trap the heat in the exhaust system, which has a direct effect on maintaining exhaust gas velocity. Clamps, brackets, and hangers are used to properly join and support the various parts of the exhaust system. These parts also help to isolate exhaust noise by preventing its transfer through the frame or body to the passenger compartment. Clamps help to secure the exhaust system parts to one another. The pipes are formed in such a way that one slips inside the other. This design makes a close fit. A U-type clamp usually holds this connection tight. Another important job of clamps and brackets is to hold pipes to the bottom of the vehicle. Clamps and brackets must be designed to allow the exhaust system to vibrate without transferring the vibrations through the car. Exhaust systems' manufacturing goes through constant changes to increase their longevity and to protect them from the harmful climate, road debris, and salts that cause rust and corrosion. Also, the exhaust system has increased outflow with reduced backpressure on the engine; this helps the engine breath, increases horsepower, and reduces emissions by the use of better catalytic converters.
Answer B is wrong. Heat shields keep the pipe temperatures more stable.
Answer C is wrong. These items are the items that hold together and support the exhaust.
Answer D is wrong. Welding of pipes is also used to connect them together.

Question #27
Answer A is wrong. Part 4 is not a catalytic converter.
Answer B is correct. Part 4 is a muffler.
Answer C is wrong. Part 4 is not a resonator.
Answer D is wrong. Part 4 is not a mini-converter.

Question #28
Answer A is correct. Only Undercar Specialist A is correct. When making a bend in the pipe, make sure the depth of the bend is exactly what is required. Then using the original pipe as a guide, rotate the pipe to position the pipe for its next bend. Work pipe from one end to the other.
Answer B is wrong. Undercar Specialist B is wrong. Do not haphazardly bend the pipe. After all of the bends have been made, it may be necessary to tweak the pipe to ensure proper placement.
Answer C is wrong. Only Undercar Specialist A is correct.
Answer D is wrong. Only Undercar Specialist A is correct.

Question #29

Answer A is correct. A restricted or clogged exhaust port in the EGR valve will not affect engine idle. Rather it will cause excessive NO_x emissions. Increases in NO_x can be caused by any condition that prevents the EGR from allowing the correct amount of exhaust gases into the cylinder or anything that allows combustion temperatures. Rough idle can be caused by a stuck open EGR valve, a PVS that fails to open, dirt on the valve seat, or loose mounting bolts (this also causes a vacuum leak and a hissing noise).

Answer B is wrong. This could also cause stalling.

Answer C is wrong. Carbon build-up will cause a sticking valve.

Answer D is wrong. Loose valve will cause a vacuum leak.

Question #30

Answer A is wrong. Undercar Specialist B is also correct.

Answer B is wrong. Undercar Specialist A is also correct.

Answer C is correct. Both undercar specialists are correct. Listening is the most common way to check for leaks. Another way, and perhaps more effective way, is to probe the system with an exhaust analyzer. With the engine running, slowly move the probe of the gas analyzer along the entire exhaust system. Before doing this, check the readings of the air in the shop. This is the base to compare against while you are probing. If, while moving the probe across the system, there is an increase in the readings, there is a leak. The key to finding the leak is remembering that it takes the analyzer approximately seven seconds to process a reading from the probe. Therefore if the readings increase, the leak is in an area you checked seven seconds ago. If you suspect an area is leaking, move the probe away from the exhaust and allow the readings to return to base. Then hold the probe at the suspected spot. If the readings begin to rise, you have found the leak. Make sure you inspect all exhaust system joints for leaks. These leaks must be corrected to protect the vehicle's occupants from being exposed to toxic gases.

Answer D is wrong. Both Undercar Specialists are correct.

Question #31

Answer A is wrong. Undercar Specialist B is also correct.

Answer B is wrong. Undercar Specialist A is also correct.

Answer C is correct. Both Undercar Specialists are correct. Worn, damaged, or broken engine/transmission mounts could be the cause of exhaust system breakage and rattles. When a technician finds an exhaust system that has broken for no apparent reason, then a thorough check of the engine and transmission mounts should be performed. Look for dried out, cracked mounts, and for loose or missing bolts. Some vehicles are prone to broken motor mounts; these vehicles will have a higher rate of exhaust system breakage from the engine's or transmission's excessive movement.

Answer D is wrong. Both Undercar Specialists are correct. When checking catalytic converter problems with a pyrometer and the temperatures are the same, you have to assume that the converter is not working at all and needs to be replaced. A plugged converter would be very hot on the inlet side and colder on the outlet side. A vacuum gauge used in the diagnosis of an exhaust excessive backpressure would show a slow response when you accelerate the engine. The needle would not go down all the way to zero on the vacuum gauge; this differs from a normal back pressure, which would go to zero quickly and come back as engine speed increased.

Question #32

Answer A is correct. After the bends are made, the pipe still needs more work before it can be installed. Once the pipe is shaped, it needs to be cut to the correct length. Before doing this, determine the needed amount of overlap for joints. Make sure the cuts are straight and even. It may be necessary to ream the inside of the pipe to remove any depression that may have resulted from the cut. The ends of an exhaust pipe must be prepared to join with other components or to serve as a tailpipe. Tools are available to properly form a flange. These flaring tools are typically built to accommodate a variety of pipe diameters. Make sure you choose the correct tools for the pipe you are working on. Also, it is wise to place any flange sealing hardware over the pipe before flaring the end. Check the flare by inserting a new gasket into the flare to make sure the flare is round and will provide a good sealing area for the gasket.

Answer B is wrong. Doing this will make the system fit together correctly.

Answer C is wrong. Pre-fitting and proper fit will make the pipes fit better.

Answer D is wrong. A warped flange may not seal correctly.

Question #33
Answer A is wrong. The inert gas the EGR puts into the cylinders lowers the combustion temperatures.
Answer B is wrong. If the EGR opens at idle or during deceleration will cause rough idle and or stalling.
Answer C is correct. Exhaust gas recirculating (EGR) systems reduce the amount of oxides of nitrogen produced during the combustion process. The EGR system dilutes the air/fuel mixture with controlled amounts of exhaust gas. Since exhaust gas does not burn, this reduces the peak combustion temperatures. At lower combustion temperatures, very little of the nitrogen in the air will combine with oxygen to form NO_x. Most of the nitrogen is simply carried out with the exhaust gases. For drivability, it is desirable to have the EGR valve opening (and the amount of gas flow) proportional to the throttle opening. Drivability is also improved by shutting off the EGR when the engine is started up cold, at idle, and at full throttle. Since the NO_x control requirements vary on different engines, there are several different systems with various controls to provide these functions.
Answer D is wrong. Some of the newer cars can control NOx with out an EGR system.

Question #34
Answer A is correct. This statement is wrong. The tailpipe carries exhaust gases and vapor out into the air, and directs them where they cannot enter the passenger compartment. The exhaust pipe, on the other hand, carries collected gases and vapor from the exhaust manifold to the next component in the exhaust system. A "Y" pipe is an exhaust pipe that connects both exhaust manifolds of a V-type engine to form a single exhaust system. An "H" pipe consists of right and left exhaust pipes connected by a balance pipe that forms a dual exhaust system. Also known as an extension or connecting pipe, the intermediate pipe connects the exhaust pipe with the muffler or resonator. Some converter systems do not have intermediate pipes.
Answer B is wrong. Y-pipes are used to make two pipes into one.
Answer C is wrong. An H-pipe has a crossover pipe to increase flow and equalize cross flow.
Answer D is wrong. The intermediate pipe is the middle pipe connecting the front and rear of the exhaust together.

Question #35
Answer A is wrong. Oxyacetylene welding is directly controlled by the welder.
Answer B is wrong. The one touch controls the flow of oxygen to the flame to increase flame temperatures.
Answer C is correct. Oxyacetylene welding provides a high-temperature flame for welding by combining oxygen and acetylene gases. The flame of this combination provides enough heat to melt most metals. Oxyacetylene welding is a manual process; the welder controls the torch movement and the welding rod. The torch is connected via hoses to separate cylinders of gas. The flow rate from the cylinders to the torch is controlled by a manifold gauge set at the tanks and by needle valves at the torch. One torch valve controls the rate of oxygen flow and the other controls the rate of acetylene flow. The two gases mix and burn at the orifice of the torch, not in the handle of the torch as stated in answer C. Oxyacetylene welding joins metals by melting and fusing. A very intense, concentrated flame is applied to the metal until a spot under the flame becomes molten and forms a pool of liquid metal. When two metals melt and the molten pools run together and fuse, the edges of the two become one. When you want to weld on the exhaust systems of cars that have been made in the last 10 years, you need to determine the types of metal from which they are made. Unlike older cars, whose exhaust systems were mostly made of mild steel tubing, the metals used today are a mixture of different types of metals. The two most common mixtures are a combination of aluminum and milled steel and stainless steel and milled steel. Both of these mixes help cut down on rust and corrosion; yet, for the technician making the repairs without some special training, they pose a problem. The gas welding of these pipes takes practice and a colder flame to keep the metal from getting brittle and breaking.
Answer D is wrong. Welding is for joining metals together.

Question #36
Answer A is correct. Only Undercar Specialist A is right. A new gasket should be used when replacing the EGR valve.
Answer B is wrong. Undercar Specialist B is wrong. Even if the gasket appears to be good, an exhaust or intake leak may occur, causing problems.
Answer C is wrong. Only Undercar Specialist A is correct.
Answer D is wrong. Only Undercar Specialist A is correct.

Question #37
Answer A is wrong. This is one of the inspections done.
Answer B is wrong. This is one of the inspections done.
Answer C is wrong. This is one of the inspections done.
Answer D is correct. To inspect a turbo unit, check the air cleaner and remove the ducting from the air cleaner to turbo and look for dirt buildup or damage from foreign objects. Check for loose clamps on the compressor outlet connections and check the engine intake system for loose bolts or leaking gaskets. Then disconnect the exhaust pipe and look for restrictions or loose material. Examine the exhaust system for cracks, loose nuts, or blown gaskets. Rotate the turbo shaft assembly. Does it rotate freely? Are there signs of rubbing or wheel impact damage? Never disassemble the turbo unit unless you have the proper equipment to reassemble it. Turbocharger replacement should only be done after the problem that caused the original one to fail has been fixed. Before any turbo is put into service, make sure it is getting the proper oil supply. Check all the oil and coolant lines for proper flow. Before starting the vehicle, disable the vehicle from starting and crank the engine until you get full oil pressure. Then, start and idle the engine for at least two minutes before driving.

Question #38
Answer A is correct. Only Undercar Specialist A is correct. Replacement catalytic converters must not only meet EPA standards, but they must also be the exact type and be an exact replacement for the original converter. When replacing a catalytic converter, always make sure that the one you are installing is a manufacturer direct or aftermarket direct replacement for the vehicle. Never alter the emissions of any vehicle.
Answer B is wrong. Converters are sized to the vehicle and model.
Answer C is wrong. Only Undercar Specialist A is correct.
Answer D is wrong. Only Undercar Specialist A is correct.

Question #39
Answer A is wrong. Undercar Specialist B is also correct.
Answer B is wrong. Undercar Specialist A is also correct.
Answer C is correct. Both Undercar Specialists are right. The federal government through the power of the EPA has established minimum requirements for the entire country. States have the right to change these standards providing these standards meet the minimum requirements set by the EPA. State and local governments that fail to comply with federal standards are subject to fines. The most important thing to know is that the new changes make it illegal for anyone to alter a vehicle's emission systems in any way. If you are asked to alter the emission control devices on a vehicle, you need to inform the person of the federal Clean Air Act. Just say No.
Answer D is wrong. Both Undercar Specialists are correct.

Question #40
Answer A is wrong. Undercar Specialist B is also correct.
Answer B is wrong. Undercar Specialist A is also correct.
Answer C is correct. Both Undercar Specialists are correct. OBD-II vehicles use a minimum of two oxygen sensors. One of these is used for feedback to the PCM for fuel control, and the other, located at the rear of the catalytic converter, gives an indication of the efficiency of the converter. If the converter is operating properly, the signal from the precatalyst O_2S will have oscillations while the post-catalyst O_2S will be relatively flat. Once the signal from the rear sensor approaches that of the front sensor, the MIL comes on and a DTC is set. When the catalytic converter is storing oxygen properly, the downstream HO_2S sensors provide low-frequency voltage signals. If the catalytic converter is not storing oxygen properly, the voltage signal frequency increases on the downstream HO_2S sensors until the frequency of the downstream HO_2S sensors approaches the frequency of the upstream HO_2S sensors. When the downstream HO_2S sensors voltage signals reach a certain frequency, a DTC is set in the PCM memory. If the fault occurs on three drive cycles, the MIL light is illuminated.
Answer D is wrong. Both Undercar Specialists are correct.

Question #41
Answer A is wrong. Undercar Specialist B is also correct.
Answer B is wrong. Undercar Specialist A is also correct.
Answer C is correct. Both Undercar Specialists are correct. Both improper pipe thickness and improper bend setups can cause bend failures. It is important that both the correct pipe sizes and equipment setup procedures be adhered to by the technician when performing bends.
Answer D is wrong. Both Undercar Specialists are correct.

Question #42
Answer A is correct. The coverage of the warranty in answer A is wrong. The performance warranty covers repairs, which are required during the first 2 years or 24,000 miles of vehicle use because the vehicle failed an emission test. Specified major emission control components are covered for the first 8 years or 80,000 miles. If the owner of the vehicle is a resident of an area with an Inspection and Maintenance (I/M) program that meets federal guidelines, the owner is eligible for this warranty protection provided that: (1) the car or light-duty truck fails an approved emissions test; (2) the vehicle is less than 2 years old and has less than 24,000 miles (up to 8 years/80,000 miles for certain components); (3) the state or local government requires that the vehicle be repaired; (4) the vehicle's test failure does not result from misuse of the vehicle or a failure to follow the manufacturer's written maintenance instructions; and (5) the owner presented the vehicle to a warranty-authorized manufacturer representative, along with evidence of the emission test failure, during the warranty period. During the first 2 years/24,000 miles, the Performance Warranty covers any repair or adjustment which is necessary to make the vehicle pass an approved, locally required emission test, and as long as the vehicle has not exceeded the warranty time or mileage limitations and has been properly maintained according to the manufacturer's specifications.
All other answers are correct. There have been many changes to the emissions warranty of vehicle manufactures over the last 30 years. As an automotive technician it is your responsibility to try and keep up with these changes; the best way is to look into the owner's manual for the emission control warranties. The EPA will show you all the new laws, but when we get an older vehicle, it only has to abide by the laws as of the date of manufacture and not after, unless a special recall has been issued to cover a defect.
Answer B is wrong. This warranty is for the base warranty period.
Answer C is wrong. In 1995 the manufacturers and the EPA were making the changes that we have today.
Answer D is wrong. There should be no cost to a customer for the repairs if due to design or defects.

Question #43
Answer A is wrong. Undercar Specialist B is also correct.
Answer B is wrong. Undercar Specialist A is also correct.
Answer C is correct. Both Undercar Specialists are correct. A heated air inlet control is used on gasoline engines with carburetion or central fuel injection. This system controls the temperature of the air on its way to the carburetor or fuel injection body. By warming the air, it reduces HC and CO emissions by improved fuel vaporization and faster warm-up. Another system uses an exhaust manifold heat control valve that routes exhaust gases to warm the intake manifold when the engine is cold. This heats the air-fuel mixture in the intake manifold and improves ventilation. The result is reduced HC and CO emissions. These control valves can be either vacuum or thermostatically operated.
Answer D is wrong. Both Undercar Specialists are correct.

Question #44

Answer A is wrong. Undercar Specialist B is also correct.
Answer B is wrong. Undercar Specialist A is also correct.
Answer C is correct. Both Undercar Specialists are correct. If a catalytic converter is found to be bad, under no circumstances should the converter be removed and replaced with a straight piece of pipe (a test pipe). This practice is illegal for the professional auto technician. Because of constant change in EPA catalytic converter removal and installation requirements, check with the manufacturer or EPA for the latest data regarding replacement. Replacement converters must have documents or labels that show it meets EPA requirements and is warranted to meet federal durability and performance standards. All manufacturers of new and rebuilt converters who meet the EPA requirements must state that fact in writing, usually in the warranty information in their catalogs. If you bypass a damaged or ineffective catalytic converter instead of replacing it or repairing it, you are breaking the law! The penalty could be much higher than the price of a new or rebuilt unit.
Answer D is wrong. Both Undercar Specialists are correct.

Question #45

Answer A is wrong. This flame will damage the metal during welding.
Answer B is correct. Let's review the three classifications of flames. A carburizing flame is blue with an orange and red end. This results from excessive acetylene. A neutral flame has a quiet blue-white inner cone. A neutral flame is the result of the correct amount of acetylene and oxygen. An oxidizing flame has a short and hissing inner cone. This flame results from an excess of oxygen in the mixture and tends to burn the metal being welded.
Answer C is wrong. This when used properly will cause a correct weld.
Answer D is wrong. When you have the correct flame it will be easier to flow the metal.

Question #46

Answer A is wrong. Undercar Specialist B is also correct.
Answer B is wrong. Undercar Specialist A is also correct.
Answer C is correct. Both Undercar Specialists are correct. The accuracy of the oxygen sensor reading can be affected by air leaks in the intake or exhaust manifold. A misfiring spark plug that allows unburned oxygen to pass into the exhaust also causes the sensor to give a false lean reading. When diagnosing the oxygen sensors, there are several things that may cause the sensor to give a false reading. First, if there is an unmetered air leak into the intake, you would get a false lean reading. If there is a leak in the exhaust system before the sensor, it would pull in unmetered air and cause a false reading. Misfiring in a cylinder would cause the oxygen sensor to see too much oxygen and cause it to see lean. In newer vehicles that have OBDII, a misfire monitor helps with this. Long-and short-term fuel monitors make diagnosing a problem easier.
Answer D is wrong. Both Undercar Specialists are correct.

Question #47

Answer A is correct. The converter cannot be replaced by anyone other than a car dealer unless it is out of warranty and a legitimate need for replacement (such as failing an emissions test, detecting damage or blockage, or a missing converter) has been established and documented. Answer A is the **EXCEPT** answer for this question. All other statements are true. Up to model year 1995, converters were covered by a 5-year/50,000-mile federal emissions warranty (7 years or 70,000 miles in California). In 1995, the warranty jumped to 8 years and 80,000 miles. You must also obtain the customer's authorization for repairs in writing; keep the paperwork for six months and the old converter for 15 days. The replacement converter must be the same type as the original and be installed in the same location.
Answer B is wrong. This is a true statement.
Answer C is wrong. This is a true statement.
Answer D is wrong. This is a true statement.

Question #48
Answer A is wrong. This is not the computer process and will turn it on the first time.
Answer B is correct. Only Undercar Specialist B is correct. The MIL will be illuminated after a fault has been detected during 2 consecutive drive cycles and a DTC will be stored in memory. If the fault does not occur during the next 40 engine warm-up cycles, after the MIL goes out, the DTC will be erased from memory. The mil light is to warn the driver that the emissions may be exceeding the EPA limits. The computer will turn this light off after a predetermined time, and only after the problem has stopped occurring. The computer puts the code in memory for later retrieval for diagnosis of the problem.
Answer C is wrong. Only Undercar Specialist B is correct.
Answer D is wrong. Only Undercar Specialist B is correct.

Question #49
Answer A is wrong. This is true.
Answer B is wrong. If the incorrect tip size is used you may have cutting problems.
Answer C is wrong. Preheating the metal will speed cutting.
Answer D is correct. All of the statements are true except answer D. Cutting oxygen exits from a central orifice of the cutting tip when the welder depresses the cutting oxygen lever. The oxyfuel gas preheating flame is adjusted in the same way as for welding. The cutting oxygen is adjusted to provide enough blow to cut but not so much that the preheating flame is blown out.

Question #50
Answer A is wrong. The springs are designed to keep the joint tight throughout all changes in temperature. When tightening these bolts, they should be tightened to the specified torque. If the bolts are tightened until the spring is compressed, there would be a rigid connection and the springs would be useless.
Answer B is wrong. The bolds are not part of the hanger.
Answer C is wrong. Neither Undercar Specialist is correct.
Answer D is correct. Neither Undercar Specialist is correct.

Question #51
Answer A is wrong. Undercar Specialist B is also correct.
Answer B is wrong. Undercar Specialist A is also correct.
Answer C is correct. Both Undercar Specialists are correct. If the sound from a turbo cycles or changes in intensity, the likely causes are a plugged air cleaner or loose material in the compressor inlet ducts of dirt buildup on the compressor wheel and housing. Most turbocharger failures are caused by one of the following reasons: lack of lubricant, ingestion of foreign objects, or contamination of lubricant. Both Undercar Specialists are right.
Answer D is wrong. Both Undercar Specialists are correct.

Question #52
Answer A is correct. Only Undercar Specialist A is correct. Common post-combustion control systems are the catalytic converter and the AIR system. The common pre-combustion control systems are the positive crankcase ventilation (PCV), engine modification systems, and exhaust gas recirculating (EGR) systems.
Answer B is wrong. Undercar Specialist B is wrong. A turbocharger is not an emission control device and therefore is not a post-combustion device.
Answer C is wrong. Only Undercar Specialist B is correct.
Answer D is wrong. Only Undercar Specialist B is correct.

Answers to the Test Questions for the Additional Test Questions Section 6

1. B
2. B
3. C
4. C
5. A
6. B
7. C
8. D
9. D
10. C
11. C
12. B
13. D
14. D
15. C
16. C
17. C
18. C
19. B
20. B
21. B
22. C
23. D
24. D
25. C
26. C
27. A
28. B
29. B
30. C
31. C
32. B
33. A
34. A
35. D
36. D
37. C
38. A
39. A
40. D
41. B
42. A
43. B
44. C
45. B
46. A
47. B
48. B
49. B
50. A
51. C
52. B
53. D
54. D
55. B

Explanations to the Answers for the Additional Test Questions Section 6

Question #1

Answer A is wrong. Also looking for carbon marks indicating leaks should be done.

Answer B is correct. Any exhaust system inspection should include listening for hissing or rumbling that would result from a leak in the system. An on-lift inspection should pinpoint any of the following types of damage: holes, road damage, separated connections, and bulging muffler seams; kinks and dents; discoloration, rust, and soft corroded metal; torn, broken, or missing hangers and clamps; and loose tailpipes or other components. A bluish or brownish catalytic converter shell indicates overheating. Answer B is not part of a visual inspection and is quite impractical. Exhaust fumes are sometimes odorless and colorless, and you may not know when the gases are getting into your passenger compartment. The first check on the system should be to listen to the exhaust for obvious leaks from the front to the back, as well as from the manifolds. Next, have someone cover the exhaust pipe while the engine is idling so that you can listen for leaks at the joints. If you still cannot find a leak, use a 4 or 5 gas analyzer to detect a leak, and also to monitor the levels of gases in the passenger compartment.

Answer C is wrong. If a disconnected pipe is found also look for damages.

Answer D is wrong. A discolored converter could indicate an internal problem and an engine problem.

Question #2

Answer A is wrong. Undercar Specialist A is wrong. The exhaust manifold gasket seals the joint between the head and exhaust manifold. On some engines there is a round gasket or doughnut that is used to connect the exhaust manifold to the exhaust pipe. This is not called the exhaust manifold gasket.

Answer B is correct. Only Undercar Specialist B is correct. On some older vehicles, there is an additional muffler, known as a resonator or silencer. This unit is designed to further reduce or change the sound level of the exhaust. It is located toward the end of the system and generally looks like a smaller, rounder version of a muffler.

Answer C is wrong. Only Undercar Specialist B is correct.

Answer D is wrong. Only Undercar Specialist B is correct.

Question #3

Answer A is wrong. Undercar Specialist A is also correct.

Answer B is wrong. Undercar Specialist B is also correct.

Answer C is correct. Both Undercar Specialists are correct. To verify that the exhaust system or converter is restricting exhaust flow, connect a vacuum gauge to an intake vacuum source. The vacuum is observed when the engine is at fast idle. If the vacuum reading decreases over time, the exhaust is restricted. Another way to check for a restricted exhaust or catalyst is to insert a pressure gauge in the exhaust manifold's bore for the O_2S. With the gauge in place, hold the engine's speed at 2000 rpm and watch the gauge. The desired pressure reading will be less than 1.25 psi. A very bad restriction will give a reading of over 2.75 psi.

Answer D is wrong. Both Undercar Specialists are correct.

Question #4
Answer A is wrong. Undercar Specialist B is also correct.
Answer B is wrong. Undercar Specialist A is also correct.
Answer C is correct. Both Undercar Specialists are correct. The air pump in a typical AIR system produces pressurized air that is sent to the exhaust manifold and to the catalytic converter. An air control valve (or air-switching valve) routes the air from the pump either to the exhaust manifold or to the catalytic converter. During engine warm-up, the valve directs the air into the exhaust manifold. Once the engine is warm, the extra air in the manifold would affect EGR operation, so the air control valve directs the air to the converter, where it aids the converter in oxidizing emissions. A thermal vacuum switch controls the vacuum to the air control valve. When the coolant is cold, it signals the valve to direct air to the exhaust manifold. Then when the engine warms to normal operating temperature, the thermal vacuum switch signals the air control valve to reroute the air to the converter. When checking the operation of the air pump and the valves in the AIR system, the flow is determined by the computer strategy for a particular drive condition. A diverter valve will direct the air flow up into the exhaust manifold during colder engine conditions to dilute the rich, unburned exhaust coming out with extra oxygen, assisting the catalytic converter to burn off the rich mixture. When the catalytic converter and the engine are up to operating temperature, the diverter valve directs the air pump flow down to the catalytic converter; this adds extra oxygen into the middle of the dual bed to facilitate burning off the hydrocarbons.
Answer D is wrong. Both Undercar Specialists are correct.

Question #5
Answer A is correct. All of the statements are correct except for answer A.
Answer B is wrong. Hangers will support the system to prevent rattles and damage.
Answer C is wrong. Spring loading the connections help to prevent breakage of the pipe.
Answer D is wrong. Heat expansion may crack the pipe.

Question #6
Answer A is wrong. Undercar Specialist A is wrong. A turbocharger depends on the engine's oil for lubrication.
Answer B is correct. Only Undercar Specialist B is correct. Undercar Specialist B is right. It is important to remember that after replacement of a turbocharger, or after an engine has been unused or stored, there can be a considerable lag after engine startup before the oil pressure is sufficient to deliver oil to the turbo's bearings. To prevent this problem, follow these simple steps: (1) Make certain that the oil inlet and drain lines are clean before connecting them. (2) Be sure the engine oil is clean and at the proper level. (3) Fill the oil filter with clean oil. (4) Leave the oil drain line disconnected at the turbo and crank the engine without starting it until oil flows out of the turbo drain port. (5) Connect the drain line, start the engine, and operate it at low idle for a few minutes before running it at higher speeds. Turbochargers are lubricated and cooled by oil supplied from the engine oil pump pressure and sometimes with engine coolant. Some use both types for additional cooling effect. The oil supply is via a small tube out of an oil galley from the engine block. To prevent bearing failure, be sure to use recommended oils and procedures when changing oils. The coolant helps to cool the oil passages in the turbo to keep the oil from being boiled and helps keep the bearings lubricated.
Answer C is wrong. Only Undercar Specialist B is correct.
Answer D is wrong. Only Undercar Specialist B is correct.

Question #7
Answer A is wrong. Undercar Specialist B is also correct.
Answer B is wrong. Undercar Specialist A is also correct.
Answer C is correct. Both Undercar Specialists are correct. Undercar Specialist A is right. A typical electronic AIR system consists of an air pump connected to a secondary air bypass (AIRB) valve, which directs the air either to the atmosphere or to the catalytic converter. Undercar Specialist B is also right. In the bypass mode, vacuum is not applied either to the valve and secondary air is vented to the atmosphere. Secondary air may be vented or bypassed due to a fuel-rich condition or during deceleration. Secondary air is also typically bypassed during cold engine cranking and cold idle conditions.
Answer D is wrong. Both Undercar Specialists are correct.

Question #8
Answer A is wrong. Neither Undercar Specialist is correct.
Answer B is wrong. Neither Undercar Specialist is correct.
Answer C is wrong. Neither Undercar Specialist is correct.
Answer D is correct. Neither Undercar Specialist is correct. The metal to be welded determines what welding method should be used. MIG can be used on all metals, whereas oxyacetylene welding is not typically used with stainless steel and must not be used on high-strength steel or aluminum. The work area is also something that needs to be considered when deciding what welding method to use. Both Undercar Specialists are wrong. Both of these types of welding have their limitations, and you can usually weld any type of exhaust system with one or the other. Oxyacetylene welding can be used on most milled steel and some mixed-metal compounds. Mig welding can be used on multiple metals as long as the correct wire is used in the gun. Some totally stainless exhaust systems may require a Tig-type welder and special welding techniques that differ from the other two types of welding mentioned.

Question #9
Answer A is wrong. Undercar Specialist A is wrong. If there are no restrictions, the vacuum reading will be high and will either stay at that reading or increase slightly as the engine runs at this speed. If the exhaust is restricted, the vacuum will begin to decrease after a period of time.
Answer B is wrong. By touching the pyrometer probe to the exhaust pipe just ahead of and just behind the converter, it is possible to read an increase of at least 100 degrees as the exhaust gases pass through the converter. If the outlet temperature is the same or lower, nothing is happening inside the converter. An exhaust restriction in the converter itself would show a higher temperature at the inlet than the outlet.
Answer C is wrong. Neither undercar specialist is correct.
Answer D is correct Neither Undercar Specialist is correct. When checking catalytic converter problems with a pyrometer and the temperatures are the same, you have to assume that the converter is not working at all and needs to be replaced. A plugged converter would be very hot on the inlet side and colder on the outlet side. A vacuum gauge used in the diagnosis of an exhaust excessive backpressure would show a slow response when you accelerate the engine. The needle would not go down all the way to zero on the vacuum gauge; this differs from a normal back pressure, which would go to zero quickly and come back as engine speed increased.

Question #10
Answer A is wrong. Undercar Specialist B is also correct.
Answer B is wrong. Undercar Specialist A is also correct.
Answer C is correct. Both Undercar Specialists are correct. Item A is a reverse-flow muffler. Item B is a straight-through muffler. The straight through design is characterized by having a straight path for the gases through a centrally located pipe with holes. Sheet metal about three times the diameter of the pipe surrounds the pipe. Most often the muffler is filled with steel wool or some other heat-resistant, sound-deadening material.
Answer D is wrong. Both Undercar Specialists are correct.

Question #11
Answer A is wrong. Undercar Specialist B is also correct.
Answer B is wrong. Undercar Specialist A is also correct.
Answer C is correct. Both Undercar Specialists are correct. OBD-II vehicles use a minimum of two oxygen sensors. One of these is used for feedback to the PCM for fuel control, and the other, located at the rear of the catalytic converter, gives an indication of the efficiency of the converter. Using a lab scope is one of the best ways to monitor or check an oxygen sensor. Also, if the converter is operating properly, the signal from the pre-catalyst O_2S will have oscillations while the post-catalyst O_2S will be relatively flat. Once the signal from the rear sensor approaches that of the front sensor, the MIL comes on and a DTC is set. When checking for proper oxygen sensor operations, start by scanning for codes. Then, watch the data stream of the sensors after the vehicle is to operating temperature. Monitor the reading for the front of the Sensor 1 locations to vary from approximately .200 volts to .800 volts or anywhere between, but quickly switching back and forth. The Sensor 2 locations are for the catalytic converter efficiency. For proper operations, this reading should stay around the .500 volts range. If any of these are out of this range, then the proper pinpoint test should performed.
Answer D is wrong. Both Undercar Specialists are correct.

Question #12
Answer A is wrong. Undercar Specialist A is wrong. The American Welding Society (AWS) recommends grade #9 or #10 for MIG welding steel.
Answer B is correct. Only Undercar Specialist B is correct. A welding filter lens, sometimes called a filter plate, is a shaded glass welding helmet insert used to protect your eyes from ultraviolet burns. The lenses are graded from 4 to 12. The higher the number, the darker the filter. Undercar Specialist B is correct. Oxyacetylene welding should be done with either a #4, #5, or #6 tinted filter shade. When welding with arc or Mig welding, use at least a shade of #12 lenses to protect your eyes from the harmful effects of the UV rays the weld arc emits. When welding with oxyacetylene, a lens of shade of #6 should be fine for most people. If these safety lenses are not of the correct levels, then you may cause irreversible damage to your eyes. When you are getting ready to weld, make sure that others around you also take the appropriate safety precautions.
Answer C is wrong. Only Undercar Specialist B is correct.
Answer D is wrong. Only Undercar Specialist B is correct.

Question #13
Answer A is wrong. The muffler B is a straight through-type muffler.
Answer B is wrong. This is a true statement.
Answer C is wrong. This type of muffler has more backpressure.
Answer D is correct. Neither of the items in the figure is best described as a staggered-flow muffler. Item A in the figure is a reverse flow muffler. Item B is a straight-through muffler. A straight-through muffler has a straight path for the gases that extends from the front to the rear of the unit. The path for the exhaust is through a centrally located pipe with holes. Sheet metal about three times the diameter of the pipe surrounds the pipe. Most often the muffler is filled with steel wool or some other heat-resistant, sound-deadening material. A reverse-flow muffler reverses the flow of the exhaust gases and has the advantage of saving space. The double shell and two shell designs are other forms of modern mufflers.

Question #14
Answer A is wrong. Restricted exhaust will cause stalling.
Answer B is wrong. Restricted exhaust will cause loss of power.
Answer C is wrong. Restricted exhaust will cause backfiring.
Answer D is correct. A restricted exhaust system can cause all of these problems. There may be many reasons for the exhaust system to become blocked. The tail pipe may have been hit, and the pipe closed off as a result. The muffler may have rusted inside and blocked the outlet to the tail pipe. The double-wall pipe may have collapsed and partially blocked the exhaust inlet to muffler. The catalytic converter may have melted and blocked all flow. The catalytic converter may have broken apart blocking pipes. The exhaust manifold may have a bad casting from the factory, with the result of blocking one total exhaust port for one cylinder.

Question #15
Answer A is wrong. Any unauthorized modification is unlawful.
Answer B is wrong. Any unauthorized modification is unlawful.
Answer C is correct. It is against the law for anyone to tamper, remove, or intentionally damage any part of the emission control system. Sometimes the law violations are obvious, such as the presence of a straight pipe where the catalytic converter should be. Some tampering is not as quickly seen, such as the enlargement of the fuel filler neck so that it can accept leaded fuels. Use of leaded fuels is deadly to the catalytic converter. Also check the piping to the AIR and EGR systems or the disconnection or plugging of control vacuum lines. If the exhaust system uses an approved catalytic converter located in the proper location, the use of a larger diameter exhaust pipe is probably not in violation of the law, unless, of course, there are state or local laws preventing this.
Answer D is wrong. Any unauthorized modification is unlawful.

Question #16
Answer A is wrong. EGR is an emission control for NOx emissions.
Answer B is wrong. PCV is the first emission device and controls combustion gases in the crankcase.
Answer C is correct. Answer C is the Environmental Protection Agency and although EPA has much to do with the reduction of exhaust emissions, it is not an emission control device.
Answer D is wrong. The EFE is for the fuel tank vapor leak feedback to the computer.

Question #17
Answer A is wrong. Undercar Specialist B is also correct.
Answer B is wrong. Undercar Specialist A is also correct.
Answer C is correct. Both Undercar Specialists are correct. There are many ways to test a catalytic converter; one of these is to simply smack the converter with a rubber mallet. If the converter rattles, it needs to be replaced and there is no need to do other testing. A rattle indicates loose catalyst substrate, which will soon rattle into small pieces. As Undercar Specialist A says, the delta temperature test uses a hand-held digital pyrometer. By touching the pyrometer probe to the exhaust pipe just ahead of and just behind the converter, there should be an increase of at least 100° F or 8 percent above the inlet temperature reading as the exhaust gases pass through the converter. Another test is the O_2 storage test which is based on the fact that a good converter stores oxygen. Undercar Specialist B is also right. Another test relies on measuring the amount of CO_2 in the exhaust. Before beginning this test, make sure the converter is warmed up. Calibrate a four- or five-gas analyzer and insert its probe into the tailpipe. Disable the ignition. Then crank the engine for 9 seconds while pumping the throttle. Watch the readings on the analyzer; the CO_2 on fuel injected vehicles should be over 11 percent and carbureted vehicles should have a reading of over 10 percent. As soon as you get your readings, reconnect the ignition and start the engine. Do this as quickly as possible to cool off the catalytic converter. If, while the engine is cranking, the HC goes above 1500 ppm, stop cranking; the converter is not working. Also stop cranking once the CO_2 readings reach 10 percent or 11 percent; the converter is good. If the catalytic converter is bad, there will be high HC and, of course, low CO_2 at the tailpipe. Do not repeat this test more than ONE time without running the engine in between.
Answer D is wrong. Both Undercar Specialists are correct.

Question #18
Answer A is wrong. Undercar Specialist B is also correct.
Answer B is wrong. Undercar Specialist A is also correct.
Answer C is correct. Both Undercar Specialists are correct. OBD-II monitors are designed to detect increased vehicle emissions with the use of upstream and downstream O_2 sensors. The upstream sensor enables the PCM to detect engine's misfires that dramatically increase exhaust oxygen content. OBD-II can also detect how efficiently the catalytic converter oxidizes by monitoring the exhaust oxygen content before and after the converter. When looking at engine misfire data, you usually do not look at an oxygen sensor output because it would not indicate which cylinder it is. It would show which engine bank is causing a problem. Use the crankshaft sensor for misfire detection. The sensor ring on the crankshaft has raised portions on it that the sensor picks up as the engine rotates, and the sensor measures the time between each raised piece on the ring. If a misfire happens, the sensor picks up a slowdown in crank speed, and the computer sees this as a misfire. The sensor wheel also has an index for cylinder # 1. So, the computer knows which cylinder misfires; if there are three consecutive cylinder misfires in a row, it turns on the check engine light.
Answer D is wrong. Both Undercar Specialists are correct.

Question #19
Answer A is wrong. A hammer may damage the pipe or component.
Answer B is correct. Use soapy solution to help install rubber insulators.
Answer C is wrong. There is no need to a special tool to install the rubber hangers.
Answer D is wrong. Heating the hanger will cause the rubber to stick.

Question #20
Answer A is wrong. This is the PCV valves purpose.
Answer B is correct. Exhaust gas recirculating (EGR) systems reduce the amount of oxides of nitrogen produced during the combustion process. The EGR system dilutes the air/fuel mixture with controlled amounts of exhaust gas. Since exhaust gas does not burn, this reduces the peak combustion temperatures. At lower combustion temperatures, very little of the nitrogen in the air will combine with oxygen to form NO_x. Most of the nitrogen is simply carried out with the exhaust gases. For drivability, it is desirable to have the EGR valve opening (and the amount of gas flow) proportional to the throttle opening. Drivability is also improved by shutting off the EGR when the engine is started up cold, at idle, and at full throttle. Since the NO_x control requirements vary on different engines, there are several different systems with various controls to provide these functions. When checking for excessive NOx, the first thing to check is the computer for codes and the proper operation of the exhaust gas recirculation valve. If both of these are working properly, check that the engine cooling system is working correctly and is at or below normal operating temperature. If all of the checks are within specifications and there are no computer codes, there may be a catalytic converter failure causing the excessive emissions.
Answer C is wrong. This is to add air to the exhaust to help the converters to burn off an extra rich condition.
Answer D is wrong. This system is to monitor the fuel system for vapor leaks.

Question #21
Answer A is wrong. This would allow unwanted air to enter the exhaust and cause the sensor to see excessive air.
Answer B is correct. All of the statements are true except answer B. A misfiring spark plug that allows unburned oxygen to pass into the exhaust causes the sensor to give a false lean (not rich) reading. The other answers are correct. The accuracy of the oxygen sensor reading can be affected by air leaks in the intake or exhaust manifold. If the HO_2S wiring, connector, or terminal is damaged, the entire oxygen sensor assembly should be replaced. Do not attempt to repair the assembly. In order for this sensor to work properly, it must have a clean air reference. The sensor receives this reference from the air that is present around the sensor's signal and heater wires. Any attempt to repair the wires, connectors, or terminals could result in the obstruction of the air reference and degraded oxygen sensor performance. Air leaking into any parts of the engine that is unmetered can affect the engine performance. The oxygen sensor readings can indicate more oxygen in the exhaust stream, and the computer would richen the mixture to the engine to compensate. Misfires can cause the oxygen sensor to think the system is too rich, and the computer will lean out that side of the engine. This may cause the other cylinders that were running good to start to falter. The oxygen sensor is very sensitive; no repairs should be attempted—you should replace it.
Answer C is wrong. On some sensors the connector and wiring is part of the reference air port.
Answer D is wrong. If the outside air reference port is blocked the sensor needs replaced.

Question #22
Answer A is wrong. You should always verify the complaint before any checks are done.
Answer B is wrong. This would only be done if you narrow the problem down to the turbo itself.
Answer C is correct. To inspect a turbocharger, start the engine and listen to the sound the turbo system makes. After listening, check the air cleaner and remove the ducting from the air cleaner to turbo and look for dirt buildup or damage from foreign objects. Check for loose clamps on the compressor outlet connections and check the engine intake system for loose bolts or leaking gaskets. Then disconnect the exhaust pipe and look for restrictions or loose material. Examine the exhaust system for cracks, loose nuts, or blown gaskets. Check for exhaust leaks in the turbine housing and related pipe connections. Visually inspect all hoses, gaskets, and tubing for proper fit, damage, and wear. Check the low pressure, or air cleaner, side of the intake system for vacuum leaks. Check all turbocharger-mounting bolts for looseness. When inspecting a turbocharger, always start with trying to duplicate the complaint first by checking the operation of the turbo in operation, then by inspecting the air filter and the inlet vanes for damage. The turbo is a precision-made component that spins at a very high rate of rpm; the balance of the vanes is very critical—any damage would cause an imbalance and destroy the turbo. The bearing is the most common failed part of a turbo; so inspection would result in a loose or binding shaft.
Answer D is wrong. If you verify that the turbo has a problem, this would be done next.

Question #23
Answer A is wrong. A restriction in the system will not effect idle.
Answer B is wrong. This test is only one of several tests to check the total EGR system.
Answer C is wrong. Neither Undercar Specialist is correct.
Answer D is correct. Neither Undercar Specialist is correct. Many technicians wrongly conclude that an EGR valve is working properly if the engine stalls or idles very rough when the EGR valve is opened. Actually this test just shows that the valve was closed and it will open. A good EGR valve opens and closes, but it also allows the correct amount of exhaust gas to enter the cylinders. If only a small amount of exhaust gas is entering the cylinder, NO_2 will still be formed. A restricted exhaust passage of only 1/8 inch will still cause the engine to run rough or stall at idle when opened, but it is not enough to control combustion chamber temperatures at higher engine speeds. Keep in mind: never assume the EGR passages are okay just because the engine stalls at idle when the EGR is fully opened. On many vehicles, the EGR passages start getting clogged right after you start driving your car. On some vehicles, the passages get clogged fast because they are pretty small (about the size of a pencil). They should be serviced on a regular basis. A lot of manufacturers do not cover this kind of service under their warranties. This passage is very important for the entire EGR system to operate efficiently.

Question #24
Answer A is wrong. Pulse air systems need no power motor to operate.
Answer B is wrong. This is normal operation to reduce backfiring in the exhaust.
Answer C is wrong. This would cause the NOx to rise.
Answer D is correct. A pulse-type AIR system uses the natural exhaust pressure pulses to pull air from the air cleaner into the exhaust manifolds and/or the catalytic converter. A manifold pipe is installed in the exhaust manifold for each engine cylinder. The inner end of these pipes is positioned near the exhaust port. The outer ends of the manifold pipes are connected to a metal container, and a one-way check valve is mounted between the outer end of each pipe and the metal container. The one-way check valves allow airflow from the metal container through the manifold pipes, but these valves prevent exhaust flow from the pipes into the container. A clean air hose is connected from the metal container to the air cleaner. At lower engine speeds, each negative pressure pulse in the exhaust manifold moves air from the air cleaner through the metal container, check valve, and manifold pipe into an exhaust port. High-pressure pulses in the exhaust manifold close the one-way check valves and prevent exhaust from entering the system. When the air is injected into the exhaust ports by the pulsed secondary air-injection system, most of the unburned hydrocarbons (HC) coming out of the exhaust ports are ignited and burned in the exhaust manifold. Since the duration of low-pressure pulses in the exhaust ports decreases with engine speed, this system is more effective in reducing HC emissions at lower engine speeds.

Question #25
Answer A is wrong. This is a true statement.
Answer B is wrong. This is a true statement.
Answer C is correct. The heated oxygen sensor monitor system monitors lean to rich and rich to lean time responses. This test can pick up a lazy O_2 sensor that cannot switch fast enough to keep proper control of the air-fuel mixture in the system. These sensors are the heated type, and the amount of time before activity of the sensor signal is present is an indication of whether it is functional or not. Some systems use current flow to indicate if the heater is working or not. All of the system's oxygen sensors are monitored once per drive cycle, but the heated oxygen sensor monitor provides separate tests for the upstream and downstream sensors. The heated oxygen sensor monitor checks the voltage signal frequency of the upstream oxygen sensors. Excessive time between signal voltage frequency indicates a faulty sensor. At certain times, the heated oxygen sensor monitor varies the fuel delivery and checks for O_2S response. A slow response in the sensor voltage signal frequency indicates a faulty sensor. The sensor signal is also monitored for excessive voltage. The heated oxygen sensor monitor also checks the frequency of the rear O_2S signals and checks these sensor signals for excessively high voltage. If the monitor does not detect signal voltage frequency within a specific range, the rear oxygen sensors are considered faulty. The heated oxygen sensor monitor will command the PCM to vary the air/fuel ratio to check the rear O_2S response.
Answer D is wrong. This is a true statement.

Question #26

Answer A is wrong. Undercar Specialist B is also correct.

Answer B is wrong. Undercar Specialist A is also correct.

Answer C is correct. Both Undercar Specialists are correct. Oxyacetylene welding begins with the proper setup of the equipment and then lighting the torch. It is important that the correct torch tip and rod be selected before proceeding. Visually inspect the condition of the equipment. If anything looks potentially hazardous, do not proceed. Turn the regulator adjusting screws all the way out before opening the cylinder valves. This prevents damage to the regulator. Stand to one side of the regulator when opening the cylinder valves. The high pressure can cause a weak or damaged regulator to burst and cause injury. Slowly open the acetylene cylinder valve one-quarter to one-half turn. Open the oxygen valve very slowly. When the regulator high-pressure gauge reaches its highest reading, turn the cylinder valve all the way open. Then open the acetylene torch valve one turn. Turn the acetylene regulator adjusting screw in slowly until the low-pressure acetylene gauge indicates a pressure that is correct for the torch tip. Then close the acetylene torch valve. Open the torch oxygen valve one turn. Turn the oxygen regulator adjusting screw in until the low-pressure oxygen gauge indicates the correct pressure for the torch tip. Then close the oxygen torch valve. Purge the lines independently by cracking open the torch valves. After purging, open the acetylene torch valve slightly. Use an igniter to light the acetylene gas at the torch's tip. Continue to open the acetylene torch valve until the flame slightly jumps away from the end of the tip. After the acetylene is regulated, slowly open the oxygen valve on the torch. As the oxygen is fed into the flame, the acetylene flame will become purple and a small inner cone will begin to form. This inner cone is white and will become whiter as more oxygen is added.

Answer D is wrong. Both Undercar Specialists are correct.

Question #27

Answer A is correct. Only Undercar Specialist A is correct. Oxygen sensors produce a voltage based on the amount of oxygen in the exhaust. Large amounts of oxygen result from lean mixtures and result in low voltage output from the O_2S. Rich mixtures release lower amounts of oxygen in the exhaust; therefore the O_2S voltage is high. The engine must be at normal operating temperature before the oxygen sensor is tested.

Answer B is wrong. Undercar Specialist B is wrong. An oxygen sensor is a voltage generator with 0 to 1 volt. There is one other oxygen sensor that is used on some newer vehicles, and it will likely replace the old style voltage-generating sensor. This sensor uses resistance to vary a voltage signal back to the computer in less startup time, about 45 seconds, rather than the 1 to 2 minutes on the older sensors. The newer sensor is called a Titania type-O_2 sensor.

Answer C is wrong. Only Undercar Specialist A is correct.

Answer D is wrong. Only Undercar Specialist A is correct.

Question #28

Answer A is wrong. Undercar Specialist A is wrong. Checking a negative back pressure EGR is like checking most other EGR valves. When vacuum is supplied to a negative back pressure EGR valve with the engine not running, the bleed port is closed, and the vacuum should open the valve.

Answer B is correct. Only Undercar Specialist B is correct. When checking EGR valves, you must understand the differences between a positive and negative, or normal, EGR valve. Most EGR valves can be checked with a hand-held vacuum pump. Typically the valve will open and hold the vacuum if the valve is good. However, if vacuum is supplied to a positive back pressure EGR valve with the engine not running, the valve will not open, because the vacuum is bled off through the bleed port. Undercar Specialist B is right. The diagnosing of EGR systems is a common task for the drivability technician that does this kind of work on a daily basis. The most common problem is carbon deposits that get in to the pintle of the valve and seat, which cause the engine to idle rough, stall or surge. When carbon gets into the EGR passageways, it can totally block them and cause the EGR system fail. The symptoms would be a pre-ignition or ping problem. This could lead to internal engine damage to the cylinders combustion chambers or pistons from excessive temperatures in the chambers. This can also cause of excessive NOx, resulting in a failed emissions test.

Answer C is wrong. Only Undercar Specialist B is correct.

Answer D is wrong. Only Undercar Specialist B is correct.

Question #29
Answer A is wrong. Some machined surfaces do not use a gasket.
Answer B is correct. Only Undercar Specialist B is correct. The figure shows an exhaust manifold and exhaust manifold gasket. The use of a gasket will help seal a warped manifold; however, some engines are originally equipped with a gasket and one must be used. If the manifold is warped beyond manufacturer's specifications or is cracked, it must be replaced.
Answer C is wrong. Only Undercar Specialist B is correct.
Answer D is wrong. Only Undercar Specialist B is correct.

Question #30
Answer A is wrong. This will prevent warpage of the welded parts.
Answer B is wrong. This if not controlled will burn through the metal.
Answer C is correct. Using MIG welding, the welder must select the size of the electrode, then set the unit to the desired voltage. The welder also must adjust the flow of the shielding gas and the rate of electrode feed. Before welding, check the equipment to make sure it is safe to use. Prior to filling in a weld, tack the parts into the position. The arc length is determined by the voltage; therefore, the welder must watch and control the distance from the nozzle to the work. By controlling the nozzle-to-work distance, the welder will control the electrode extension distance. The welding speed and torch angle determines the bead width and appearance.
Answer D is wrong. Also the voltage setting makes a difference.

Question #31
Answer A is wrong. Undercar Specialist B is also correct.
Answer B is wrong. Undercar Specialist A is also correct.
Answer C is correct. Both undercar specialists are correct. A converter outlet gasket should never be reused. Replace any exhaust joint with a broken gasket or seal, or one which is suspected of leaking.
Answer D is wrong. Both Undercar Specialists are correct.

Question #32
Answer A is wrong. This reading cannot go over the emission levels for the year of the vehicle standards.
Answer B is correct. Only Technician B is correct. OBD-II systems are designed to turn on the MIL when the vehicle has any failure that could potentially cause the vehicle to exceed its designed emission standard by a factor of 1.5 (not 2.5). The system does that by the use of various monitors. If one or more monitored systems are out of compliance, then the MIL turns on to indicate a problem. If the vehicle is experiencing a malfunction that may cause damage to the catalytic converter, the MIL will flash once per second. If the MIL flashes, the driver needs to get the vehicle serviced immediately. If the driver reduces speed or load and the MIL stops flashing, a code is set and the MIL stays on. This means the conditions that presented potential problems to the converter have passed with the changing operating conditions. While we are discussing the OBDll system operating strategy, let's take a look at the systems that turn on the check engine light. Two of the most important things are the fuel control and the catalytic converter systems. These two systems control the amount of fuel that goes into the engine. The OBDII monitors itself through the use of oxygen sensors in the exhaust to get feedback, indicating whether it's giving to little or too much fuel to the engine to limit the emissions.
Answer C is wrong. Only Technician B is correct.
Answer D is wrong. Only Technician B is correct.

Question #33

Answer A is correct. The results from welding with oxyacetylene are determined by the mixture of the gases. Certain conditions are necessary for a good weld. The temperature of the flame must be great enough to melt the metals. There must also be enough heat to overcome any heat losses. The flame must not add dirt, foreign material, or carbon to the metal being worked. The flame must not burn or oxidize the metal. The products of combustion should not be toxic. The amount of heat is determined by the amount of gas being burned. To obtain more heat, a torch tip with a larger orifice is used. To provide the right amount of gas to the larger orifice, higher pressures are set. Likewise, if low heat is needed, a torch tip with a smaller orifice is used with lower pressures. Regardless of the torch tip or the pressures used, if the flame is neutral it will have the same temperature regardless. More heat is present with a larger tip because the flame covers more area at one time. If the torch tip orifice is too small, not enough heat will be available to bring the metal to its melting and flowing temperature. If the torch tip is too large, poor welds will result because the weld will need to be made too quickly, the welding rod will melt too quickly, and the weld pool will be hard to control. When using the gas torch for welding, there are a few things to remember. Look at the metal you are going to weld and choose the proper size tip for the thickness of the metal. Start the torch with the gas; then, slowly add oxygen until you get a pinpoint blue flame and no excessive hissing from the tip. To weld most exhaust steel tubing, the flame should be at the proper heat and mixture.

Answer B is wrong. This is true and the tip size would increase for metal thickness.

Answer C is wrong. A bigger tip should be used for the thicker metal.

Answer D is wrong. A tip too large will cause overheating and the tip should be reduced.

Question #34

Answer A is correct. Only Undercar Specialist A is correct. An air bypass valve (or diverter valve) diverts, or detours, air during deceleration. Excess air in an exhaust rich with fuel can produce a backfire or explosion in a muffler. The HCs in the exhaust can burn with the air from the air pump, causing a backfire. To prevent this, the air from the air pump is quickly diverted to the atmosphere. One-way check valves allow air into the exhaust but prevent exhaust from entering the pump in the event the drive belt breaks. The valve opens to let air in but closes to keep exhaust from leaking out. The check valve can be checked with an exhaust gas analyzer. Start the engine and hold the probe of the exhaust gas analyzer near the check valve port. If any amount of CO or CO_2 is read, the valve leaks. If this valve is leaking, hot exhaust is also leaking which could ruin the other components in the air-injection system. When diagnosing a vehicle that is backfiring through the exhaust system, look at several things. First, is there an engine drivability complaint, such as running lean, rich, or anything that could contribute to this complaint? Next, check the entire exhaust system for any leaks that may be letting any outside air into the exhaust stream. Look at the AIR system for proper operation. The most likely cause from this system is that the air bypass or diverter valve is unable to dump the air pump air to outside during deceleration. Check the valve for a bad or leaking diaphragm.

Answer B is wrong. A bad valve will allow air to go the wrong direction and introduce air to the flow.

Answer C is wrong. Only Undercar Specialist A is correct.

Answer D is wrong. Only Undercar Specialist A is correct.

Question #35

Answer A is wrong. Undercar Specialist A is wrong. During acceleration the EGR valve is closed. Therefore if it did not open, it would be working normally and would not have an adverse affect on acceleration.

Answer B is wrong. For this same reason as A; Undercar Specialist B is also wrong.

Answer C is wrong. Neither Undercar Specialist is correct.

Answer D is correct. Neither Undercar Specialist is correct. For this same reason, Undercar Specialist B is wrong.

Question #36
Answer A is wrong. This reading will be between .1 and 1.0 volts and the normal operating readings will float between .4 and .7.
Answer B is wrong. This reading will be between .1 and 1.0 volts and the normal operating readings will float between .4 and .7.
Answer C is wrong. Neither Undercar Specialist is correct.
Answer D is correct. Neither Undercar Specialist is correct. An oxygen sensor's voltage should cycle from low voltage to high voltage. The signal from most oxygen sensors varies between 0 and 1 volt. If the voltage is continually high, the air/fuel ratio may be rich or the sensor may be contaminated. When the O_2S voltage is continually low, the air/fuel ratio may be lean, the sensor may be defective, or the wire between the sensor and the computer may have a high-resistance problem. If the O_2S voltage signal remains in a mid-range position, the computer may be in open loop or the sensor may be defective.

Question #37
Answer A is wrong. This is the easiest way to bend a new pipe without the card method.
Answer B is wrong. This method is preferred for pre-making a pipe.
Answer C is correct. Having the correct bends is not enough; the bends must be at the right place and at the right angle.
Answer D is wrong. If you go too far on a bend it is hard to go back without damage to the pipe.

Question #38
Answer A is correct. Only Undercar Specialist A is correct. Oxygen sensors produce a voltage based on the amount of oxygen in the exhaust. When measuring exhaust gases with an analyzer, use one that tests 5 different gases: HC, CO, CO_2, O_2, and NO_x. When diagnosing a problem with the performance of an engine, look at all of these gases. The HC (hydrocarbons) are raw unburned fuel; CO (carbon monoxide) is partly burned fuel and particulates; CO_2 (carbon dioxide) is a byproduct of clean combustion; O_2 (oxygen) is the air we breathe; and NO_x (oxides of nitrogen) are from excessive combustion chamber temperatures (above about 2500 degrees).
Answer B is wrong. Undercar Specialist B is wrong. Large amounts of oxygen result from lean mixtures and result in low voltage output from the O_2S. Rich mixtures release lower amounts of oxygen in the exhaust; therefore the O_2S voltage is high.
Answer C is wrong. Only Undercar Specialist A is correct.
Answer D is wrong. Only Undercar Specialist A is correct.

Question #39
Answer A is correct. In MIG welding, the polarity of the power source is important in determining the penetration to the workpiece. Direct current (DC) power sources used for MIG welding typically use DC reverse polarity. DC reverse polarity means the wire (electrode) is positive, and the workpiece is negative. Weld penetration is greater using this connection. Weld penetration is also greatest using CO_2 gas. However, CO_2 gives a harsher, more unstable arc, which leads to increased spatter. So when welding on thin materials, it is preferable to use argon/carbon dioxide. Voltage adjustment and wire feed speed must be set according to the diameter of the wire being used and metal thickness.
Answer B is wrong. This gas will make the weld penetrate better.
Answer C is wrong. This gas keeps the oxygen out of the weld.
Answer D is wrong. For the correct weld the speed is also important.

Question #40
Answer A is wrong. Mini-converter/preheater is used close to the manifold for lowering emissions during warm-up.
Answer B is wrong. Leaded fuels will damage all converters.
Answer C is wrong. EGR systems are to recirculate exhaust gases to reduce NOx.
Answer D is correct. A mini-converter/preheater is used to reduce emissions during engine warm-up. Because a converter is most efficient when it is warmed up and the regular converters take some time to heat up, a small converter is installed in or at the exhaust manifold. This mini-converter heats up quickly and begins to work well before the regular converters.

Question #41

Answer A is wrong. Undercar Specialist A is wrong. Using an air chisel on the bolt would be a desperate attempt. The air chisel would likely damage the flanges, as well as the bolt. If the flange is destroyed, the pipe assembly would need to be replaced and there would be no need to loosen the bolt. The assembly could be just cut off.

Answer B is correct. Only Undercar Specialist B is correct. A great way to loosen a frozen bolt is to try to tighten it first. That will break up the dirt and rust that is keeping the bolt from loosening.

Answer C is wrong. Only Undercar Specialist B is correct.

Answer D is wrong. Only Undercar Specialist B is correct.

Question #42

Answer A is correct. Only Undercar Specialist A is correct. The emissions warranties apply to used vehicles, as well as new ones, as long as the vehicle has not exceeded the warranty time or mileage limitations.

Answer B is wrong. If a vehicle has had all the proper maintenance performed and it still fails an emissions test, the catalytic converter should be checked, not randomly replaced. There are many other possible causes for emission test failures. The cause for the failure should be found and corrected.

Answer C is wrong. Only Undercar Specialist A is correct.

Answer D is wrong. Only Undercar Specialist A is correct.

Question #43

Answer A is wrong. Continued driving may cause engine damage from carbon fouling.

Answer B is correct. If a misfire that threatens engine or catalyst damage occurs, the misfire monitor flashes the MIL on the first occurrence of the misfire. A fault detected by the catalyst monitor must occur on three drive cycles before the MIL is illuminated. For the misfire and fuel system monitors, if the fault does not occur on three consecutive drive cycles under similar conditions, the MIL is turned off. For the catalyst efficiency, HO_2S, EGR, and comprehensive component monitors, the MIL is turned off if the same fault does not reappear for three consecutive drive cycles. When the fault is no longer present and the MIL is turned off, the DTC is erased after 40 engine warm-up cycles. A technician may use a scan tester to erase DTCs immediately. A pending DTC is a code representing a fault that has occurred but that has not occurred enough times to illuminate the MIL.

Answer C is wrong. This is if the problem no longer appears or has been repaired.

Answer D is wrong. This is so if a problem does appear the old code is no longer present.

Question #44

Answer A is wrong. Undercar Specialist B is also correct.

Answer B is wrong. Undercar Specialist A is also correct.

Answer C is correct. Both Undercar Specialists are correct. Worn, damaged, or broken engine/transmission mounts could be the cause of exhaust system breakage and rattles. A quick check of the mounts can be done by pulling up and pushing down on the engine or transmission while watching the mount. If the mount's rubber separates from the metal plate or if the engine or transmission case moves up but not down, replace the mount. If there is movement between the metal plate and its attaching point on the frame, tighten the attaching bolts to the appropriate torque and recheck. If it is necessary to replace the mounts, make sure you follow the procedure for maintaining the alignment of the drive line. Failure to do this may result in poor transmission operation and excessive exhaust system rattles. Checking engine and transmission mounts should start with checking for fluid leaks on and around the mounts. Next, check for obvious breaks and torn mounts. Next, put the vehicle in gear; have a helper to power brake the engine a little at a time in forward and reverse gears to see if there is excessive lifting of the engine or transmission movement indicating broken mounts. Fix any broken mounts before any repair attempts to the exhaust.

Answer D is wrong. Both Undercar Specialists are correct.

Question #45
Answer A is wrong. This type of valve will not open at idle due to the bleed port.
Answer B is correct. Only Undercar Specialist B is correct. The positive back pressure EGR valve has a bleed port and valve positioned in the center of the diaphragm. A light spring holds this bleed valve open, and an exhaust passage is connected from the lower end of the tapered valve through the stem to the bleed valve. When the engine is running, exhaust pressure is applied to the bleed valve. At low engine speeds, exhaust pressure is not high enough to close the bleed valve. If control vacuum is supplied to the diaphragm chamber, the vacuum is bled off through the bleed port, and the valve remains closed. Considering this, you can see that Undercar Specialist A is wrong. Until the exhaust pressure is great enough to close the bleed valve, all vacuum is bled off and the valve will not open. Undercar Specialist B is correct; the vacuum would bleed off and the diaphragm would not move. As engine and vehicle speed increase, the exhaust pressure also increases. At a preset throttle opening, the exhaust pressure closes the EGR valve bleed port. When control vacuum is supplied to the diaphragm, the diaphragm and valve are lifted upward, and the valve is open. If vacuum from an external source is supplied to a positive back pressure EGR valve with the engine not running, the valve will not open because the vacuum is bled off through the bleed port.
Answer C is wrong. Only Undercar Specialist B is correct.
Answer D is wrong. Only Undercar Specialist B is correct.

Question #46
Answer A is correct. Only Undercar Specialist A is correct. The activity of the sensor can be monitored on a scanner. By watching the scanner while the engine is running, the O_2S voltage should move to nearly 1 volt then drop back to close to zero volts. Immediately after it drops, the voltage signal should move back up. This immediate cycling is an important function of an O_2S. If the response is slow, the sensor is lazy and should be replaced. With the engine at about 2500 rpm, the O_2S should cycle from high to low 10 to 40 times in 10 seconds.
Answer B is wrong. Undercar Specialist B is wrong. When testing the O_2S, make sure the sensor is heated and the system is in closed loop.
Answer C is wrong. Only Undercar Specialist A is correct.
Answer D is wrong. Only Undercar Specialist A is correct.

Question #47
Answer A is wrong. This would increase hydrocarbons and oil contamination.
Answer B is correct. The exhaust gas recirculation (EGR) system introduces exhaust gases into the intake air to reduce the temperatures reached during combustion. This reduces the chances of formingNO_x during combustion. The cars that are covered under the laws that govern the NO_x levels are not the only cars that produce this toxic gas. When the car manufactures had to start leaning out the cars engines to pass the newest EPA standards, they started to increase the combustion temperatures to levels above 2500 degrees. They didn't realize that the process was producing a gas that is more toxic that the HC and COs that they were trying to reduce. Once they realized this, they came up with the exhaust gas recirculation valve, which started to send exhaust gases back into the engine to help cool the combustion temperatures to lower levels.
Answer C is wrong. Air itself would not effect NOx unless too hot of air is getting to the engine.
Answer D is wrong. This system would affect NOx.

Question #48
Answer A is wrong. If this is not done failure of the new turbo will result.
Answer B is correct. Never add more than the proper amount of oil to an engine. This can cause oil consumption and leaks. After replacement of a turbocharger, or after an engine has been unused or stored, there can be a considerable lag after engine start up before the oil pressure is sufficient to deliver oil to the turbo's bearings. To prevent this problem, follow these simple steps: (1) When installing a new or remanufactured turbocharger, make certain that the oil inlet and drain lines are clean before connecting them. (2) Be sure the engine oil is clean and at the proper level. (3) Fill the oil filter with clean oil. (4) Leave the oil drain line disconnected at the turbo and crank the engine without starting it until oil flows out of the turbo drain port. (5) Connect the drain line, start the engine, and operate it at low idle for a few minutes before running it at higher speeds. (6) Turn off the engine and make sure the engine has the correct oil level.
Answer C is wrong. This makes sure flow is coming thru the turbo.
Answer D is wrong. This is to make sure oil has lubricated the turbo properly.

Question #49
Answer A is wrong. A high pitch noise can be damage to the bearings or fins.
Answer B is correct. Only Undercar Specialist B is correct. To inspect a turbocharger, start the engine and listen to the sound the turbo system makes. As a technician becomes more familiar with this characteristic sound, it will be easier to identify an air leak between the compressor outlet and engine or an exhaust leak between engine and turbo by the presence of a higher pitched sound. If the turbo sound cycles or changes in intensity, the likely causes are a plugged air cleaner, loose material in the compressor inlet ducts, or dirt buildup on the compressor wheel and housing. When testing a turbo operation, look at the entire intake and exhaust systems, including the air filter. The turbo's biggest enemy is dirt and debris from either a dirty air filter or no air filter. The turbo turns at speeds around 10,000 rpms, and even little pieces of dirt can damage the turbine blades. The turbo also has bearings that support the turbine shaft, These are both water and oil cooled; so, you will have to also look at both of these systems.
Answer C is wrong. Only Undercar Specialist B is correct.
Answer D is wrong. Only Undercar Specialist B is correct.

Question #50
Answer A is correct. Part 2 is a catalytic converter.
Answer B is wrong. Part 2 is not a muffler.
Answer C is wrong. Part 2 is not a resonator.
Answer D is wrong. Part 2 is not a mini-converter/preheater.

Question #51
Answer A is wrong. This is the first inspection done.
Answer B is wrong. This will show if the weld holds under pressure.
Answer C is correct. All of the methods listed, except a vacuum test, are valid methods that can be used to verify the integrity of a weld. A vacuum test on an assembled exhaust system could not be done.
Answer D is wrong. The noise test with a hose or an electronic tester can detect small leaks.

Question #52
Answer A is wrong. Overloading and off road driving can damage the converter.
Answer B is correct. The use of high-octane fuels may be recommended by the manufacturer and therefore would not be in violation of the law. Also, burning high-octane fuel in any engine would not hinder the effectiveness of the engine and therefore would not be considered misuse or lack of proper maintenance. Examples of misuse and lack of proper maintenance include the following: vehicle abuse such as off-road driving or overloading; tampering with emission control parts or systems, including removal or intentional damage of such parts or systems; and improper maintenance, including failure to follow maintenance schedules and instructions specified by manufacturer, or use of replacement parts which are not equivalent to the originally installed parts.
Answer C is wrong. This is illegal and could get a fine and the only repair would be to put the vehicle back to original condition.
Answer D is wrong. Maintenance is very important and could cause premature component failures.

Question #53
Answer A is wrong. A pipe cutter is a common item used to make a clean cut to a pipe.
Answer B is wrong. A torch is a common item used to make a clean cut to a pipe.
Answer C is wrong. When using a plasma arc cutter make sure you disconnect the battery.
Answer D is correct. Metal Inert Gas (MIG) is a method of welding, not cutting. Manual pipe cutters can be used to totally cut through an old pipe as can heat. Oxyacetylene equipment should not be used on high-strength steel components for welding or cutting. Plasma arc cutting is replacing oxyacetylene as the best way to cut metals. Plasma arc cutting cuts damage metal effectively and quickly but does not destroy the properties of the base metal.

Question #54
Answer A is wrong. Part 1 is nor a catalytic converter.
Answer B is wrong. Part 1 is not a muffler.
Answer C is wrong. Part 1 is not a resonator.
Answer D is correct. Part 1 is an oxygen sensor.

Question #55
Answer A is wrong. The sensor wire and connector are part of the vent for the sensor.
Answer B is correct. All of the statements are true except for answer B. Never apply contact cleaner or other materials to the sensor or wiring harness connectors. These materials may get into the sensor, causing poor performance. When you have determined that the oxygen sensor is in need of replacement, you should make sure that the exhaust is cool to the touch so that the threads will not come out of the pipe with the sensor. Make sure a high temperature anti-seize is used on the threads of the new sensor. Then, after putting in hand tight, then tighten an extra 1⁄4 to 1⁄2 turn—do not overtighten, or sensor damage will result.
Answer C is wrong. This is true.
Answer D is wrong. This should only be done if the repair manual says during testing.

Glossary

Air cleaner A device connected to the carburetor in a manner that all incoming air must pass through it. Its purpose is to filter dirt and dust from the air before it passes into the engine.

Air filter A filter that removes dust, dirt, and particles from the air passing through it.

Air/fuel (A/F) ratio The ratio, by weight, of air to gasoline entering the intake in a gasoline engine. The ideal ratio for complete combustion is 14.7 parts of gasoline to 1 part of fuel (i.e., stoichiometry).

Attainment area A geographic area in which levels of a criteria air pollutant meet the health-based primary standard (national ambient air quality standard, or NAAQS) for the pollutant. An area may have an acceptable level for one criteria air pollutant but may have unacceptable levels for others. Thus, an area could be both attainment and nonattainment at the same time. Attainment areas are defined using federal pollutant limits set by EPA.

Battery A device for storing energy in chemical form so it can be released as electricity.

Bearing Soft metallic shells used to reduce friction created by rotational forces.

California Air Resources Board (CARB) An agency established in 1967 by the California legislature to enforce its own emissions standards for new vehicles based on California's unique need for more stringent controls.

Carbon monoxide (CO) A colorless, odorless, poisonous gas, produced by incomplete burning of carbon-based fuels, including gasoline, oil, and wood. Carbon monoxide is also produced from incomplete combustion of many natural and synthetic products. For instance, cigarette smoke contains carbon monoxide. When carbon monoxide gets into the body, the carbon monoxide combines with chemicals in the blood and prevents the blood from bringing oxygen to cells, tissues, and organs. The body's parts need oxygen for energy, so high-level exposures to carbon monoxide can cause serious health effects, with death possible from massive exposures. Symptoms of exposure to carbon monoxide can include vision problems, reduced alertness, and general reduction in mental and physical functions. Carbon monoxide exposures are especially harmful to people with heart, lung, and circulatory system diseases.

Catalyst A material that promotes a chemical reaction without itself being changed by the reaction. The noble metals platinum, palladium, and rhodium are used as a catalyst in catalytic converters.

Catalytic converter An emission device located in front of the muffler in the exhaust system. It looks very much like a heavy muffler and contains catalysts to clean up an engine's emissions before they leave the end of the exhaust pipe.

Clean Air Act The original Clean Air Act was passed in 1963, but our national air pollution control program is actually based on the 1970 version of the law. The 1990 Clean Air Act Amendments are the most far-reaching revisions of the 1970 law. In this summary, we refer to the 1990 amendments as the 1990 Clean Air Act.

Combustion Burning. Many important pollutants, such as sulfur dioxide, nitrogen oxides, and particulates (PM-10) are combustion products, often products of the burning of fuels such as coal, oil, gas, and wood.

Diagnostic trouble codes (DTCs) Codes associated with engine controller fault messages that identify emission control components that are malfunctioning and can be retrieved using a diagnostic scan tool.

Direct current (DC) A type of electrical power used in mobile applications. A unidirectional current of substantially constant value.

EGR The exhaust gas recirculation system that allows a small amount of exhaust gas to be routed into the incoming air/fuel mixture to reduce NO_x emissions.

Electronic fuel injection (EFI) A generic term applied to various types of fuel injection systems.

Emissions Gases and particles left over after the combustion event of an engine or from a fuel system. The primary emissions of concern are hydrocarbons (HC), carbon monoxide (CO), and oxides of nitrogen (NO_x).

Environmental Protection Agency (EPA) The federal government agency, created as a result of the 1970 Clean Air Act (CAA), that establishes regulations and oversees the enforcement of laws related to the environment. Included in these laws are regulations on the amount and content of automotive emissions and for onboard diagnostic systems.

Exhaust manifold A component which collects and then directs engine exhaust gases from the cylinders.

Exhaust valve An engine part that controls the expulsion of spent gases and emissions out of the cylinder.

Federal Test Procedure (FTP) A transient-speed mass emissions test conducted on a loaded dynamometer. This is the test that, by law, car manufacturers use to certify that new vehicles are in compliance for hydrocarbon, carbon monoxide, and oxides of nitrogen emissions and must be passed before that model may be sold in the United States.

Filler neck Tube fitted to the fuel tank that allows fuel to be added from a remote location.

Fuel filter A device located in the fuel line to remove impurities from the fuel before it enters the carburetor or injector system.

Fuel injection Injecting fuel into the engine under pressure.

Fuel pump A mechanical or electrical device used to move fuel from the fuel tank to the carburetor or injectors.

Gasket A thin layer of material or composition that is placed between two machined surfaces to provide a leakproof seal between them.

Heat shield An assembly designed to restrict the transfer of high heat from one component to another.

Heated oxygen sensor (HO_2S) An oxygen sensor that is heated electrically as well as by engine exhaust so that it warms to normal operating temperature more quickly.

Hydrocarbon (HC) A family of organic fuels containing only hydrogen and carbon. Gasoline consists almost entirely of a hydrocarbon mixture, and high levels of hydrocarbons in tailpipe emissions are a result of unburned gasoline.

Inspection and maintenance program (I/M program) Auto inspection programs are required for some polluted areas. These periodic inspections, usually done once a year or once every two years, check whether a car is being maintained to keep pollution down and whether emission control systems are working properly. Vehicles that do not pass inspection must be repaired. As of 1992, 111 urban areas in 35 states already had I/M programs. Under the 1990 Clean Air Act, some especially polluted areas will have to have enhanced inspection and maintenance programs using special machines that can check for such things as how much pollution a car produces during actual driving conditions.

Intermediate pipe Part of the exhaust system, primarily used to connect different major parts of the system together.

Malfunction indicator light (MIL) The instrument panel light used by the OBD-II system to notify the vehicle operator of an emissions-related fault. The MIL is also known as the "service engine soon" or "check engine" lamp.

Misfire When incomplete or no combustion occurs in one or more cylinders due to improper fuel, ignition, cylinder compression, or air.

Monitor In the context of OBD-II, monitor refers to any onboard diagnostic test that is executed by the PCM. A monitor is simply a diagnostic test run on a component that determines if it is operating within design specifications.

Muffler (1) A hollow, tubular device used in the lines of some air conditioners to minimize the compressor noise or surges transmitted to the inside of the car. (2) A device in the exhaust system used to reduce noise.

Nitrogen oxides (NO_x or oxides of Nitrogen NO_x) Harmful gaseous emissions; gases that form when nitrogen from the air is combined with oxygen under conditions of high temperature and pressure in the combustion chamber. Oxides of nitrogen contribute to the formation of ground-level ozone and photochemical smog.

On-Board Diagnostics (OBD) A program that assesses the condition of the emission system, including the sensors and the computer itself and communicates its findings to the technician by means of diagnostics trouble codes.

Open loop A state in which the air/fuel mixture is being controlled by the engine computer according to a standard program and essentially ignores the oxygen sensor signal; normally encountered during the first few minutes of operation after a cold start.

Oxygen Sensor (O_2S) An input sensor that sends a voltage signal to the computer in relation to the amount of oxygen in the exhaust stream.

PCM Power train control module.

Pollutants (pollution) Unwanted chemicals or other materials found in the air. Pollutants can harm health, the environment, and property. Many air pollutants occur as gases or vapors, but some are very tiny solid particles such as dust, smoke, or soot.

Power train control module (PCM) The onboard control module that monitors engine functions or both engine and transmission/transaxle functions.

Resonator A second muffler in-line with the other muffler.

Scan Tool A hand-held computer that is plugged into a vehicle's data link connector allowing the technician to read diagnostic trouble codes, readiness status, freeze frame data, and other information.

Seal Generally refers to a compressor shaft oil seal; matching shaft-mounted seal face and front head-mounted seal seat to prevent refrigerant and/or oil from escaping. May also refer to any gasket or O-ring used between two mating surfaces for the same purpose.

Secondary Air Injection (AIR) An emissions system found primarily on large-engine vehicles that pumps fresh air into the exhaust stream to reduce HC and CO emissions.

Sensor An electrical device used to provide a computer with input information about engine operating parameters.

Smog A mixture of pollutants, principally ground-level ozone, produced by chemical reactions in the air involving smog-forming chemicals. A major portion of smog-formers, volatile organic compounds, are found in products such as paints and solvents. Smog can harm health, damage the environment, and cause poor visibility. Major smog occurrences are often linked to heavy motor vehicle traffic, sunshine, high temperatures, and calm winds or temperature inversion (weather condition in which warm air is trapped close to the ground instead of rising). Smog is often worse away from the source of the smog-forming chemicals, since the chemical reactions that result in smog occur in the sky while the reacting chemicals are being blown away from their sources by winds.

Tailpipe The part of the exhaust system that allows the exhaust gases to leave the rear of the vehicle and be emitted into the atmosphere.

Turbocharger A small radial fan pump driven by the energy of the exhaust flow.

Notes

Notes

Notes

Notes

Notes

Notes